CELL PHYSIOLOGY

ENCYCLOPAEDIA OF CELL BIOLOGY - II

CELL PHYSIOLOGY

CELL PHYSIOLOGY

By

Dr. M. Prakash

Dept. of Zoology

M.M.H. Post Graduate College

Ghaziabad (U.P.)

(India)

DISCOVERY PUBLISHING HOUSE PVT. LTD.

NEW DELHI-110 002

First Published - 2010

Reprinted - 2018

ISBN: 978-81-8356-454-0 (Set)

ISBN: 978-81-8356-560-8

Cell Physiology

Published by:

DISCOVERY PUBLISHING HOUSE PVT. LTD.

4383/4B, Ansari Road, Darya Ganj

New Delhi-110 002 (India)

Phone: +91-11-23279245, 43596064-65

Fax: +91-11-23253475

E-mail: discoverypublishinghouse@gmail.com

sales@discoverypublishinggroup.com

web: www.discoverypublishinggroup.com

Printed at:

Infinity Imaging Systems

Delhi

Preface

The present title "Encyclopaedia of Cell Biology" has been written for undergraduate and postgraduate students of all Indian universities. It is a text on structural and functional unit of the organism that has been thoroughly checked and ornamented with easy and clear illustrations. Special emphasis has been concentrated on the fundamentals of the subject. All related students will find this well established text eminently suitable for introductory courses. Professional cell biologists, researchers, and students may find it useful for updating, and reference purposes. One of our major concerns of this title is to provide students with a basic understanding of what a cell is and why the cell is the fundamental unit of life. Our approach throughout has been to focus on the major question involved and the experimental approaches utilized in addressing these questions. Each important point is illustrated with an example to help clarify the idea; the numerous line drawings have all been chosen or created to help convey particular points. The result is a text that is easy to read, without resorting to verbal gimmicks or talking down to readers.

To make the work more comprehensive and informative, the author has consulted many authoritative books, research journals, abstracts, monographs etc., so there can be no claim to originality except in the manner of treatment.

The author expresses his thanks to his friends and colleagues whose continue inspirations have initiated him to bring out this book.

The author expresses his gratitude to Mr. Wasan and staff of M/s Discovery Publishing House Pvt. Ltd. for their whole hearted cooperation in the publication of this book.

Author

Contents

1

INTRODUCTION

Cell physiology is the study of life activities of cells-plant cells, animal cells and microorganisms. The study of these activities has been a very vital one in the past few years. The present era is to the biologist what the beginning of the twentieth century was to the physicist, for never before in the history of biology has so much been achieved in so short a time toward an understanding of the fundamental problems of the living cell.

The life activities of the cell might be enumerated as follows:

(1) nutrition, including in its broadest sense the uptake and preparation of food and oxygen for use in the cells, the release of energy in cells and the elimination of the wastes produced,

(2) response to the environment,

(3) growth and

(4) reproduction or cell division, and differentiation.

The briefest comparison of plants, animals and microorganisms serves to show the fundamental similarities of their functions despite differences in structure. The body of a seed plant is composed of cells organized into a number of complex organs: root, stem, leaf, flower and, in the mature plant, the fruit with the developing seed. Through the root hairs water and nutrient salts pass into the vascular tissue and are carried through the stem to the leaves. Carbon dioxide enters through the stomata of the leaves, and during the light of day carbohydrates are synthesized from carbon dioxide and water in the chlorophyll-bearing cells. From the carbohydrates and salts the plant later synthesizes other compounds such as amino acids, fatty acids and watersoluble vitamins. These manufactured foodstuffs are then distributed in a solution (the sap) to cells in all parts of the plant, and the excess may be stored in the seed, fruit, root, stem or, less frequently, leaf. The vascular system and the surrounding woody fibers serve as a skeleton supporting the entire plant.

Although photosynthesis occurs only in some special cells of the seed plant, all cells of the plant body must be supplied with cell foods, and they must be freed of cell wastes. Plants grow, using part of the food manufactured, and after reaching maturity they reproduce, using more of it. Growing plants respond to gravity, to light and to chemicals and other factors in their environment. Thus the root grows downward in response to gravity, whereas the stem grows upward; the root grows away from the light while the stem grows toward it (for instance, everyone has seen the sunflower follow the sun). The response of plants to tactile stimuli is seen in the twining tendrils which fasten some plants to a support. Movements of plants are mainly growth movements, and although they are too slow for us to see, they can be readily demonstrated in timelapse photography. The more rapid turgor movements can be seen in the sensitive plant *Mimosa*.

The essential architectural plan of a vertebrate, a triploblastic form (embryonically three-layered), consists of an outer tube, the body wall, enclosing an inner tube, the gut, and supported by an internal skeleton consisting of bones. The head constitutes the anterior end and contains many of the sense organs or receptors. The trunk, consisting of a body wall and the internal organs in it, is supported by the appendages. The internal organs are organized into various organ systems. For example, the organs of the digestive system prepare food for use in the body, and the digested food is absorbed into the blood and is distributed over the entire body by the organs of the circulatory system. As in the plant, regardless of the organ in which it is found, the individual cell requires the same essential cell foods. The respiratory system takes up the oxygen which is distributed to all cells by way of the circulatory system, and it voids to the air the carbon dioxide brought to it by the circulatory system. The blood flushes wastes from cells in all parts of the body, carrying these wastes to the kidneys where they are extracted from the blood and voided to the exterior. The animal may be considered to have developed a set of services of supply-digestive, respiratory, circulatory and excretory-which maintain the cellular environment. The needed supplies of oxygen and food are brought to the cells, and the wastes of cellular activity are removed from the cells and voided to the exterior. By way of receptor organs (such as the eyes and ears) sensitive to light, sound, pressure, gravity, heat or chemicals, the animal receives information about the environment. Response to the information so gathered is made possible by a nervous and muscular system supported by a bony skeletal system. After reaching sexual maturity the animal reproduces.

As can be seen, the same fundamental activities of life-nutrition, response, growth and reproduction-are therefore performed by plants and animals, though in somewhat different ways.

However, complex organs like those found in higher plants and animals are not necessary for the performance of the fundamental biological functions in all organisms. Respiration takes place in bacteria, protozoa and unicellular plants, all without a respiratory system; likewise, digestion can occur without the aid of a digestive system and excretion, without the aid of an excretory system. Microorganisms get food and oxygen directly from the environment and void their wastes directly to it. They respond to the environment, and if conditions are favorable, they grow and reproduce.

The common denominator required for the performance of these fundamental activities of plants, animals and microorganisms is clearly the cell, not the complex of organs and systems which has developed in multicellular plants and animals. These systems are concerned primarily with maintaining the internal environment in contact %ith the cells of the organisms. It is through the cell of the multicellular animal and plant that all the flow of matter and flux of energy occurs. The cell is a fundamental unit, not only of structure but also of function of all living things.

VITALISTIC AND MECHANISTIC INTERPRETATION OF LIFE ACTIVITIES

The mind of man demands an explanation for the phenomena of nature. Primitive peoples developed mythologies which conceived of the earth and man as being controlled by various supernatural forces. Such forces were later suggested to explain biological phenomena. Accounts of these early conceptions may be found in various histories of biology.

As information was accumulated from observation and experimentation, it eventually became clear that the grosser functions in the animal and plant, such as digestion, respiration, circulation and excretion, are subject to the laws. of physics and chemistry and some theorists, the "*mechanists*," think that all the phenomena of life are thus explainable. Others, the "*vitalists*," maintain that although digestion, respiration, excretion and circulation might be interpreted in terms of chemical and physical laws, more complex processes such as the orderly development of the embryo and the complex mental life of man can be interpreted best in terms of something apart from and in addition to the physical-chemical laws. They postulate a vital force, categorically equivalent to matter and energy, to perform this role.

The cell physiologist finds it profitable to operate on the working hypothesis that the physiological activities of the cell are the consequence of the operation of a physical-chemical system which constitutes the cell. If the functions of the cell are the consequences of the operation of a physical-chemical system performing work, then the cell may be likened to a machine. It is the purpose of the cell physiologist, first, to determine the nature of the machinery of the cell and, second, to determine the principles which govern its operation, with the hope of better understanding its functioning. For this reason the book begins with a consideration of thermodynamics and energy flow in biological systems. To do its work the cell must liberate chemical energy from cell foods and couple the energy-liberating reactions with energy-requiring reactions. The mechanistic hypothesis has been fruitful in stimulating experimentation and accumulating useful information.

EXPERIMENTAL APPROACH IN CELL PHYSIOLOGY

Cell physiology employs the experimental method, many examples of which will be given in later chapters. The solution to a problem in science is often achieved by slow and laborious collection of data. Pertinent observations are made and data already available are focused on the problem in question and a possible explanation, a hypothesis, is formulated. Carefully planned experiments are then prepared for testing it, and it may happen that after much work an investigator finds that the attempted explanation is inadequate and that he has only cleared the ground for further analytical work. An alternative hypothesis is then constructed and tested. Occasionally decisive experiments result in dramatic advances in knowledge and understanding. Such decisive experiments have occurred more frequently in physics and chemistry, in which the total number of variables is fewer. In the biological system, in which there are a great number of variables, similar progress in theoretical knowledge may be expected only as the physiologist gains control of these factors.

The recent analysis of the structure of proteins and nucleic acids and the deciphering of the genetic code are considered by some to be such advances in biology. These findings have created a revolution in biological thought such as had not occurred since the concept of evolution had stirred the biological world. Although it is difficult to evaluate a matter so close at hand, the birth of a whole new field called *molecular biology* resulting from these discoveries and its rapid growth attest to the significance of the findings.

Experimentation makes possible selection of the most likely explanation for a phenomenon. As the unlikely explanations are eliminated, the remaining one becomes more probable, but still subject to revision as more decisive data become available.

By the use of the experimental method, the following cellular problems have been studied:

1. The nature of the organization of cells of plants, animals and micro organisms as a key to an under standing of cellular activities.
2. The nature of the cell environment and the resistance and adjustment of the cell to the variations of the environment.
3. The nature of the cell membrane and the regulation of transport of materials into and out of the cell across the membrane.
4. The nature of cell foods and their interconversions and the mechanism by which the process of respiration releases energy from cell foods.
5. Use of the energy liberated in respiration for performing the various types of work done by the cell, e.g., for the maintenance of irritability or readiness to respond, for motility required for the responses, for osmotic work and for the manufacture of secretions and of new constituents leading to cell growth and cell division.

A consideration of each of these problems in turn constitutes the major portion of this book. This approach leads from the problems of organization of the cell to cellular dynamics. It also leads from the more descriptive aspects of cell physiology to the more complex studies of respiration, irritability, growth and cell division. It has the advantage that much of the descriptive information is already available when the more difficult problems are analyzed. At the end, a brief chapter on the history of cell physiology brings into perspective some of the major trends of research in the field.

THE RELATION OF CELL PHYSIOLOGY OF VARIOUS FILELDS OF PHYSIOLOGY

Much of *animal physiology* deals with the work of the various organs of the body and with the coordination of the various organs in the integrated behavior of the animal as a whole. Since the animal physiologist is concerned with functions of organs his study might be called organ physiology. Medical men in the past have been primarily concerned with vertebrate organ physiology, especially of mammals, because the information gained can be more directly applied to the physiology of man in health and disease. Therefore, medical physiology most often meant mammalian organ physiology. However, to interpret organ physiology, research workers in medical physiology have often found it necessary to work at the cellular level as well. Therefore much overlapping occurs at the present time.

Comparative animal physiology is a study of organ functions in various types of animals, vertebrate and invertebrate, in an attempt to find fundamental relations not recognized when the function is studied in only one group of animals. The different ways in which each function may be' performed in various animals are therefore viewed in the perspective of the whole animal kingdom.

Plant physiology concerns itself with the response, nutrition, growth and reproduction of various types of plants. Because many of these problems first required solution at the cell level, plant physiologists have contributed greatly to the development of cell physiology.

Since functioning of animals and plants depends ultimately upon the functioning of their component cells, an understanding of cell physiology is basic to all the other physiologi!al studies. This has been recognized by researchers, many of whom, interested primarily in the higher organisms, have worked at the cell level as a background for solution of problems at the organ level.

THE LITERATURE OF CELL PHYSIOLOGY

The most recent achievements in cell physiology and data on current studies are presented at scientific meetings and are recorded in current issues of such periodicals as *Science*, *Nature*, the *Journal of General Physiology*, *The Journal of Cellular and Comparative Physiology*, the *Journal of Experimental Biology*, *Experimental Cell Research, Physiological Zoology, Biochimica et Biophysica Acta, Proceedings of the National Academy of Sciences, U. S.*, *Archiv fur die gesampte Physiologie, Compte Rendu de la Societe de Biologie, Doklady Academii Nauk SSSR* (Biological Sciences Division), *Biofizika* and many others. Some of the Russian journals are now available in English translation.

Reviews or summaries of recent advances on subjects of interest to the cell physiologist appear in such journals as the *Quarterly Review of Biology, Physiological Reviews, Biological Reviews of the Cambridge Philosophical Society, Annual Review of Physiology,* annual reviews in a large number of specialized fields in cell physiology, symposia on special subjects (e.g., permeability, metabolism), and others. Some reviews will be cited in the subsequent chapters.

There are many good monographs on various special subjects of interest to the cell physiologist, e.g., on protoplasmic structure, permeability, oxidation-reduction and respiratory enzymes, bioluminescence, muscle chemistry and physiology. In this book reference will be made to such monographs when they are pertinent to the subject matter being discussed in a given chapter.

Compendia and textbooks cover much larger fields than monographs, and consequently their treatment of specific topics is usually less complete. Reference is given at the end of this chapter to textbooks of this kind which have been found most useful. Most of them will not be specifically cited again in this book, but they are fruitful sources of additional material, especially in the fields designated.

If a bibliography is to be compiled on a subject of interest, journals which abstract periodicals are helpful. The most useful abstract journals for the cell physiologist are *Biological Abstracts* and *Chemical Abstracts, International Abstracts of Biological Sciences, Berichte uber die gesampte Physiologie, Berichte uber die Wissenschaftliche Biologie, Inhaltsverzeichnisse sowjetscher Fachzeitschr lien: III B, Chemie and Biologie, B.A.S.LC., Current Contents,* and *Bulletin de l'Institut Pasteur.*

These journals not only give references to recent original papers, reviews and monographs, but they also furnish abstracts of the contents.

Included in the references at the end of each chapter are general references it which the reader may find additional in formation covering the general field of th(chapter, usually from a different point o: view. Citations to other literature support views expressed on subjects of lesser scope They are included as an aid to those whc may wish to get a more complete treat ment of some topic which interests hirr especially.

2

CHEMICAL ENVIRONMENT OF CELL

Despite their great diversity, the chemical composition and metabolic processes of all living things are remarkably similar. At one time it was believed that there was a unique substance in their chemical composition that distinguished living organisms from inanimate matter. The search for this substance proved fruitless. Instead, biologists came to understand that it is the specific organization and precise interaction of its components that characterize life. To understand life processes, then, one must know the basic principles of chemistry, or, to be more specific, biochemistry. For those who have never enjoyed a formal chemistry course, in other chapters we provide the basic definitions and concepts necessary for an understanding of biology. Those who have already studied chemistry may find them a useful review.

CHEMICAL ELEMENTS

All matter, living and nonliving alike, is composed of chemical *elements*, substances that cannot be broken down into simpler substances by chemical reactions. The matter of the universe is composed of 92 naturally occurring elements, ranging from hydrogen, the lightest, to uranium, the heaviest. In addition to the naturally occurring elements, about 17 elements heavier than uranium have been made by bombarding elements with subatomic particles in devices known as particle accelerators.

About 98% of an organism's mass' is composed of only six

elements-oxygen, carbon, hydrogen, nitrogen, calcium, and phosphorus. Approximately 14 other elements are consistently present in living things, but in smaller quantities. Some of these, such as iodine and copper, are known as *trace elements* because they are present in such minute amounts. Table elsewhere in this chapter lists the elements that make up a living organism and explains why each is important.

Irystead of writing out the name of each element, chemists use a system of abbreviations called *chemical symbols-usually* the first one or two letters of the English or Latin name of the element. For example, O is the symbol for oxygen, C for carbon, Cl for chlorine, N for nitrogen, and Na for sodium (its Latin name is *natrium)*. Chemical symbols for the elements found in living organisms are given in Table elsewhere in this chapter.

THE ATOM

Imagine a bit of gold being divided into smaller and smaller pieces. The smallest possible particle of gold that could be obtained would be an atom of gold. The *atom* is the smallest subdivision of an element that retains the characteristic chemical properties of that element. The subdivision of any kind of matter ultimately yields atoms. This is true no matter what physical state matter may assume-solid, liquid, or gas. Atoms are almost unimaginably small-much smaller than the tiniest particle visible under a light microscope. By special scanning electron microscopy, with magnification as much as 5 million times, researchers have been able to photograph some of the larger atoms, such as uranium.

Atomic Structure

An atom is composed of smaller components called *subatomic particles*. For our purposes we need consider only three types-protons, neutrons, and electrons. *Protons* have a positive electric charge; *neutrons* are uncharged particles with about the same mass as protons. Protons and neutrons make up almost all the mass of an atom and are concentrated in the *atomic nucleus*. *Electrons* have a negative electrical charge and an extremely small mass (only about 1/1800 of the mass of a proton). The electrons, as we will see, spin about in the space surrounding the atomic nucleus.

Each kind of element has a fixed number of protons in the atomic nucleus. This number, called the *atomic number*, is written as a subscript to the left of the chemical symbol. Thus $_{1}H$ and $_{8}0$ indicate that the hydrogen nucleus contains one proton and the oxygen nucleus has eight

Table 2.1 : Elements that make up the human body.

Name	Chemical symbol	Approximate composition by mass (%)	Importance or function
Oxygen	O	65	Required for cellular respir-ation; present in most organic compounds; component of water
Carbon	C	18	Backbone of organic molec-ules;, can form fuur,bonds-with other atoms
Hydrogen	H	10	Present in most organic compound; component of water
Nitrogen	N	3	Component of all proteins and nucleic acids
Calcium	Ca	1.5	Structural component of bones and teeth; important in muscle contraction, conduction of nerve impulses, and blood Clotting
Phosphorus	P	1	Component of nucleic acids; structural component of bone; important in energy transfer Potassium K 0_4 Principal positive ion-(cutipn) within cells; important in nerve function; affects muscle contraction
Sulfur	S	0.3	Component of most proteins
Sodium	Na	0.2	Principal positive ion in-interstitial (tissue) fluid; important In fluid balance; essential for conduc-tion of nerve impulses
Magnesium	Mg	0.1	Needed in blood and' ocher body tissues
Chlorine	Cl	0.1	Principal negative ion (anion) of interstitial fluid; important in fluid balance; component of sodium chloride
Iron	Fe	Trace amount	Component of hemoglobin and myoglobin; component of certain enzymes
Iodine	1	Trace amount	Component of thyroid hormones

Other elements found in very small amounts in the body include manganese (Mn), copper (Cu), zinc (Zn), cobalt (Co), fluorine (F), molybdenum (Mo), selenium (Se), and a few oth-ers. They are called trace element$, protons. It is the atomic number, the number of protons in the nucleus, that determines the chemical identity of the atom. The total number of protons plus neutrons in the nucleus is termed the *mass number* and is indicated by a superscript to the left of the chemical symbol. The common form of oxygen atom, with eight protons and eight neutrons in its nucleus, has an atomic number of 8, and a mass number of 16. It is indicated by the symbol $^{16}_{8}O$.

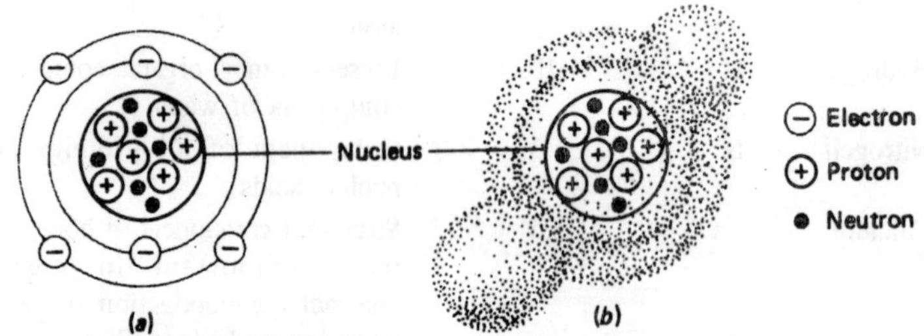

Figure 2.1 : Two ways of representing an atom. (a) Bohr model of an atom. Although the Bohr model is not an accurate way to depict electron configuration, it is commonly used because of its simplicity and convenience. (b) An electron cloud. Dots represent the probability of the electron's being in that particular location at any given moment.

ELECTRON CONFIGURATION

The way electrons are arranged around an atomic nucleus is referred to as the atom's *electron configuration*. Knowing the loacations of electrons enables chemists to predict how atoms can combine to form different types of chemical compounds.

Becuase they are negatively charged, electrons are attracted to the positively charged protons in the atomic nucelus (opposite charges tend to attract). At the same time, electrons repel one another (like charges tend to repel one another). These consderations help determione the loactions of electrons.

An atom may have several **energy levels, or electron shells,** where electrons are located. The lowest energy level is the one closest to the nucleus. Only two electrons can occupy this energy level. The second energy level can accommodate a maximum of eight electrons. Although the third and outer shells can each contain more than eight electrons, they are most stable when only eight are present. We may consider the

first shelf complete when it contains *two electrons, and every other shell* complete when it contains eight electrons. The atomic structures of some elements that are important in biological systems-carbon, hydrogen, oxygen, nitrogen, sodium, and chlorine-are shown in Figure elsewheere in this chapter.

Although the simple diagrams, called *Bohr models*, of electron configuration shown in Figure elsewheere in this chapter are helpful in understanding atomic structure, they are highly oversimplified. Within energy levels, electrons occur in characteristic regions of space, termed *orbitals*. There may be several orbitals within a given energy level, and each orbital can contain at most two electrons. Electron orbitals may be represented by spherical, dumbbell-shaped, or more complex three-dimensional coordinates.

Electrons are thought to whirl around the nucleus, now close to it, now farther away. Orbitals represent the places where electrons are most probably found. In fact, one way of illustrating an atom is

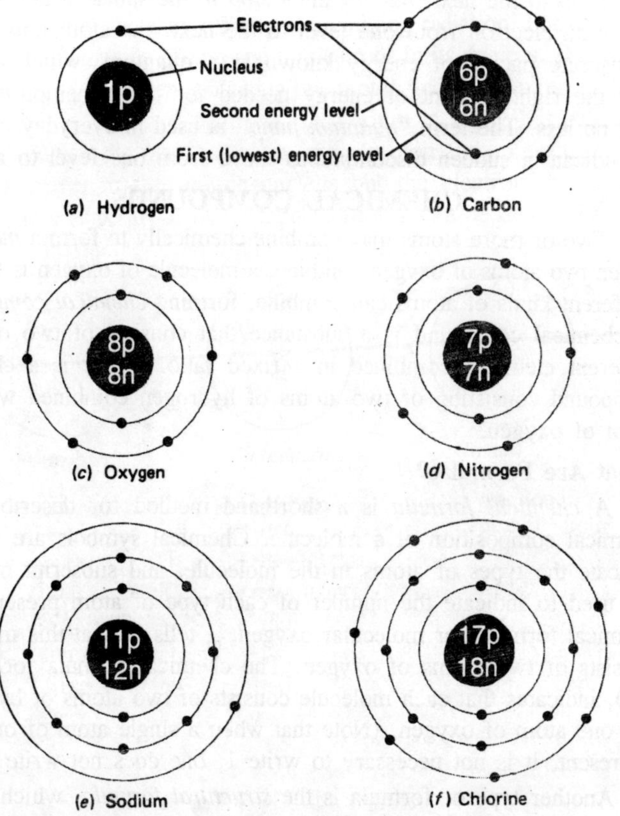

Figure 2.2 : Bohr models of some biologically important atoms.

to show its orbitals as electron clouds. Each occupied orbital can be represented as an electron cloud. The density of the shaded areas represents the likelihood that an electron is present at a given time.

The distance of the electrons from the atomic nucleus depends upon their respective energy levels. The specific orbital that an electron occupies is determined by the total energy of the system and the number of electrons available to the atom. An electron can be moved to an orbital farther from the nucleus by providing it with more energy, or an electron can give up energy and sink back to a lower energy level in an orbital nearer the central nucleus. Energy is required to move a negatively charged electron further away from the positively charged nucleus.

When energy is added to the system, an electron can jump from one level to the next, *but it* cannot *stop in the space in between.* To move an electron from one level to the next, the atom must absorb a discrete packet of energy known as a quantum, which contains just the right amount of energy needed for the transition-no more and no less. The term "*quantum jump*" is used in everyday language to indicate a sudden discontinuous move from one level to another.

CHEMICAL COMPOUNDS

Two or more atoms may combine chemically to form a *molecule*. When two atoms of oxygen combine, a molecule of oxygen is formed. Different kinds of atoms can combine, forming *chemical compounds*. A chemical compound is a substance that consists of two or more different elements combined in a fixed ratio. Water is a chemical compound consisting of two atoms of hydrogen combined with one atom of oxygen.

What Are Formulas?

A *chemical formula* is a shorthand method for describing the chemical composition of a molecule. Chemical symbols are used to indicate the types of atoms in the molecule, and subscript numbers are used to indicate the number of each type of atom present. The chemical formula for molecular oxygen, $_{02,}$ tells us that this molecule consists of two atoms of oxygen. The chemical formula for water, H_2O, indicates that each molecule consists of two atoms of hydrogen and one atom of oxygen. (Note that when a single atom of one type is present, it is not necessary to write 1; one does not write H_2O_1.)

Another type of formula is the *structural formula*, which shows not only the types and numbers of atoms in a molecule but also their arrangement. In any type of chemical compound the atoms are always arranged in the same way. From the chemical formula for water, H_2O, you could only guess whether the atoms were arranged H-H-O or H-O-H. The structural formula, H-O-H, settles the matter, indicating that the two hydrogen atoms are attached to the oxygen atom. In water this arrangement is the only one chemically possible. However, there are other substances that consist of the same atoms yet have different chemical properties due to alternative arrangements. Such compounds are known as *structural isomers*. The sugars glucose and fructose are examples.

Chemical Equations

During any moment in the life of an organism, be it an earthworm or a pine tree, many complex, highly organized chemical reactions are taking place. The chemical reactions that occur between atoms and moleculesfor instance, between methane (natural gas) and oxygen-can be described on paper by means of chemical equations:

$$\underset{\text{Methane}}{CH_4} + \underset{\text{Oxygen}}{2O_2} \rightarrow \underset{\text{Carbon dioxide}}{CO_2} + \underset{\text{Water}}{2H_2O}$$

In a chemical equation the reactants (the substances that participate in the reaction) are written on the left side of the equation and the products (the substances formed by the reaction) are written on the right side. The arrow means yields and indicates the direction in which the reaction tends to proceed. The number preceding a chemical symbol or formula indicates the number of atoms or molecules reacting. Thus 202 means two molecules of oxygen and 2H20 means two molecules of water. The absence of a number indicates that only one atom or molecule is present.

In some cases the reaction will proceed in the reverse direction as well as forward; at equilibrium a certain amount of the product continuously breaks up to form the reactants, and the rate of the forward reaction equals the rate of the reverse action. Reversible reactions are indicated by double arrows:

$$\underset{\text{Nitrogen}}{N_2} + \underset{\text{Hydrogen}}{3H_2} \rightleftarrows \underset{\text{Ammonia}}{2NH_3}$$

HOW ATOMS COMBINE: CHEMICAL BONDS

The chemical properties of an element are determined primarily by the number and arrangement of electrons in the outermost energy level (electron shell). In a few elements, called the noble gases, the

Table 2.2 : Some biologically important ions.

Name	Formula	Charge
Sodium	Na^+	1 +
Potassium	K^+	1 +
Hydrogen	H^+	1 +
Magnesium	Mg^{2+}	2 +
Calcium	Ca^{2+}	2 +
Iron	Fe^{2+} or Fe^{3+}	2 + [iron(II)] or 3+ [iron(1H)]
Ammonium	NH_4^+	1 +
Chloride	Cl^-	1 –
Iodide	I^-	1 –
Carbonate	CO_3^{2-}	2 –
Bicarbonate	HCO_3^-	1 –
Phosphate	PO_4^{3-}	3 –
Acetate	CH_3COO^-	1 –
Sulfate	SO_4^{2-}	2 –
Hydroxide	OH^-	1 –
Nitrate	NO_3^-	1 –
Nitrite	NO_2^-	1 –

outermost shell is filled. These elements are chemically inert, meaning that they will not readily combine with other elements. Two such elements are helium with two. electrons (a complete inner shell) and neon with ten electrons (a complete inner shell of two and a complete second shell of eight). The electrons in the outermost energy level of an atom are referred to as valence electrons. When the outer shell of an atom contains fewer than eight electrons, the atom tends to lose, gain, or share electrons to achieve an outer shell of eight (zero or two in the case of the lightest elements).

The elements in a given compound are always present in a certain proportion by mass. This reflects the fact that atoms are attached to each other by chemical bonds in a precise way to form the compound. A chemical bond is the attractive force that holds two atoms together. Each bond represents a certain amount of potential chemical energy. The atoms of each element form a specific number of bonds with the atoms of other elements-a number dictated by the number of valence electrons. The two principal types of chemical bonds are covalent bonds and ionic bonds.

Covalent Bonds

Covalent bonds involve the sharing of electrons between atoms. The more precise definition of a molecule is a combination of two or more atoms joined by covalent chemical bonds. A simple example of a covalent bond is the one joining two hydrogen atoms in a molecule of hydrogen gas, H_2. Each atom of hydrogen has one electron, but two electrons are required to complete the first energy level. Each hydrogen atom has the same capacity to attract electrons, so one does not donate an electron to the other. Instead, the two hydrogen atoms share their single electrons so that each of the two electrons is attracted simultaneously to the two protons in the two hydrogen nuclei. The two electrons are thus under the influence of *both* atomic nuclei, and they join the two atoms together.

The carbon atom has four electrons in its outer energy level. These four electrons are available for covalent bonding $-\overset{|}{\underset{|}{C}}-$. When one carbon and four hydrogen atoms share electrons a molecule of methane (CH_4) is formed:

```
    H
    |
H—C—H
    |
    H
```

Each atom shares its outer-level electrons with the others, thereby completing the first energy level of each hydrogen atom and the energy level of the carbon atom. When one electron pair is shared between two atoms, the covalent bond is referred to as a single bond.

The nitrogen atom has five electrons in its outer shell. When a nitrogen atom shares electrons with three hydrogen atoms, a molecule of ammonia, NH_3, is formed:

```
H—N—H
    |
    H
```

Two oxygen atoms may achieve stability by forming covalent bonds with one another. Each oxygen atom has six electrons in its outer shell. To become stable, the two atoms share two pairs of electrons, forming molecular oxygen. When two pairs of electrons are shared in this way, the covalent bond is referred to as a *double bond*. Some atoms form triple bonds with one another, sharing three pairs of electrons.

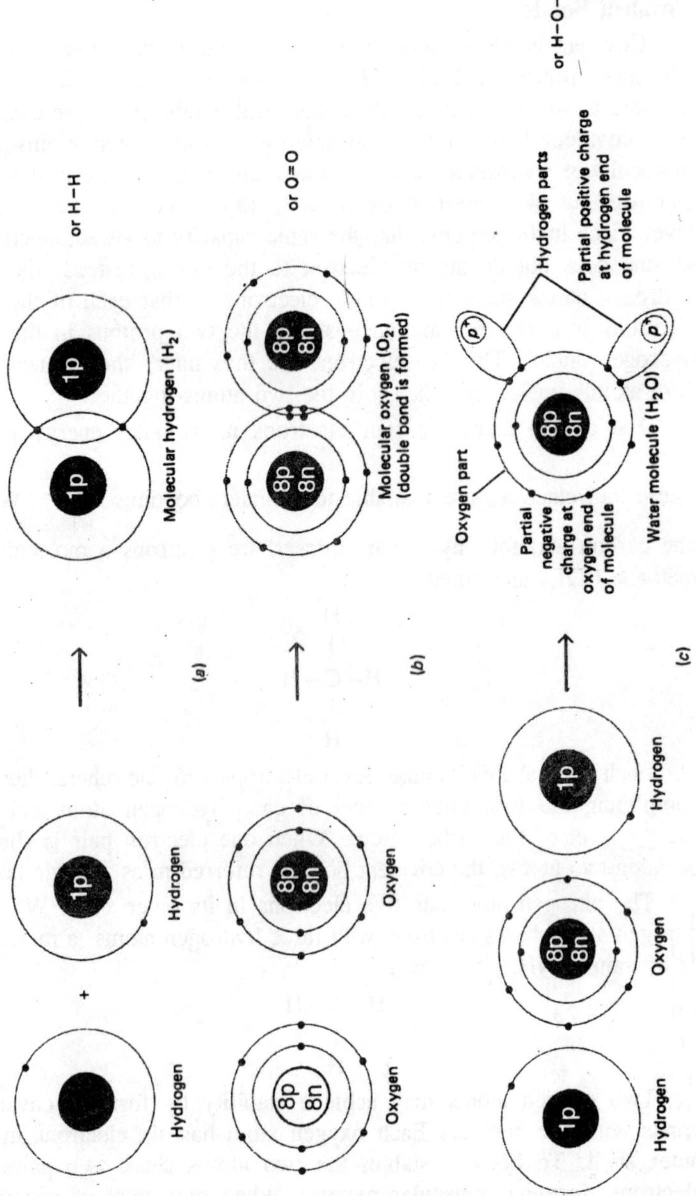

Figure 2.3 : Formation of covalent compounds.

Electronegativity is a measure of an atom's attraction for the electrons in chemical bonds. A covalent bond between atoms of different electronegativities is known as a *polar covalent bond*. The polarity of compounds is very important in understanding the structure of biological membranes and their properties. Covalent bonds have many degrees of polarity, from *nonpolar*, in which the electrons are exactly shared (as in the hydrogen molecule), to those in which the electrons are *much* closer to one atom than to the other.

Ionic Bonds

One extreme of polarity is the ionic bond, in which electrons are pulled completely from one atom to the other. The number of protons in the nucleus remains unchanged, so the loss or gain of electrons produces an atom with a net positive or negative charge. Such electrically charged atoms are termed ions. Atoms with one, two, or three electrons in the outer shell generally donate electrons to other atoms. These atoms, then, become positively charged because of the excess of protons in the nucleus. For example, sodium has one valence electron. It tends to donate this electron, becoming a sodium ion (Na^+) with a one plus charge. Calcium tends to lose its two valence electrons to become a calcium ion (Ca^{2+}) with a two plus charge. Positively charged ions are termed cation.

Atoms with five, six, or seven electrons in their outer shell tend to *gain* electrons from other atoms and become negatively charged anions (e.g., Cl^-, chloride ion). Charged particles, both anions and cations, play many important roles in biological systems, such as the transmission of nerve impulses and the contraction of muscles.

An *ionic bond* is the force of electrical attraction between two oppositely charged ions. When held together by ionic bonds, oppositely charged ions form an ionic compound. Sodium chloride is a good example of an ionic compound. A sodium atom, with atomic number 11, has two electrons in its inner shell, eight in the second, and one in the third shell. A sodium atom cannot fill its third shell by obtaining seven electrons from other atoms because it would then have a vast excess of negative charge. Instead, it gives up the single electron in its third shell to some electron acceptor, leaving the second shell as the complete outer shell. A chlorine atom, atomic number 17, has two electrons in its inner shell, eight in the second, and seven in the third shell. The chlorine atom achieves a complete outer shell not by losing the seven electrons in its third shell (for it would then have a vast positive charge) but by accepting an electron from an electron donor, such as sodium, to complete its third shell.

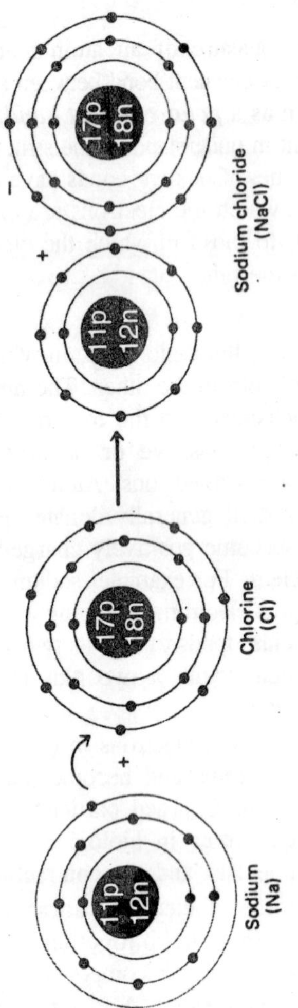

Figure 2.4 : Formation of an ionic compound. Sodium donates its single valence electron to chlorine, which has seven electrons in its outer energy levels.

When an electron donor, such as sodium, meets an electron acceptor, such as chlorine, the electron may be transferred completely from the donor to the acceptor. The sodium ion now has 11 protons in its nucleus and ten electrons around the nucleus, giving it a net charge of 1^+. The chlorine ion has 17 protons in its nucleus and 18 electrons around the nucleus, so it has a net charge of 1^-. These ions attract each other as a result of their opposite charges. They are held together by this electrical attraction to form sodium chloride (NaCl), common table salt. This transfer of one or more electrons from one atom to another and the binding together of two ions of opposite charge (the ionic bond) results in the formation of an ionic compound. Ionic bonds occur between electron donors and electron acceptors.

Whether an ionic compound is in solid form or is dissolved in water, its constituent particles (ions) do not share electrons. Because of this, the term *molecule* does not adequately explain the properties of ionic compounds such as NaCl. Chemists simply refer to them as compounds.

Hydrogen Bonds

Another type of bond that is important in biological systems is the *hydrogen bond*. This type of bond is formed when a hydrogen atom is attracted to two other atoms, one of which is usually oxygen or nitrogen. Hydrogen bonds tend to form between a hydrogen atom covalently bonded to oxygen or nitrogen and some other electronegative atom, usually oxygen or nitrogen. The atoms involved may be in two parts of the same molecule or in two molecules. When hydrogen is

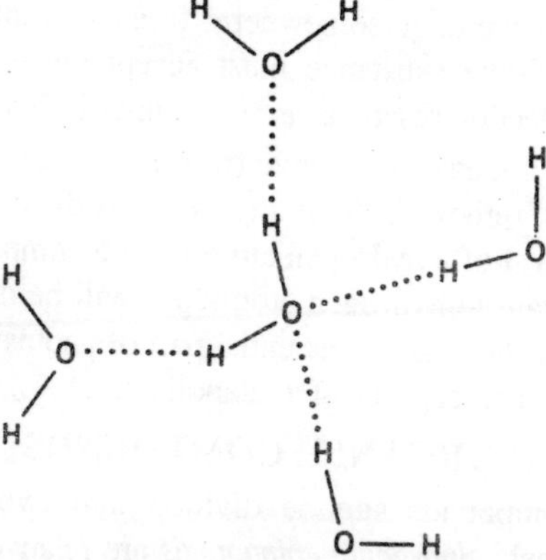

Figure 2.5 : Hydrogen bonding of water molecules. Each water molecule tends to form hydrogen bonds with four neighbouring water molecules.

combined with an electronegative atom, such as oxygen, it has a partial positive charge because its electron is positioned closer to the oxygen atom. Hydrogen bonds are weak and are readily formed and broken. They have a specific length and orientation, which is very important in their role in helping to determine the three-dimensional structure of large molecules, such as nucleic acids and proteins. Though relatively weak individually, the large numbers of hydrogen bonds that occur in the double helix of DNA and in the alpha helix of proteins are together very effective.

OXIDATION-REDUCTION

Rusting-the combination of iron with oxygen-is a familiar example of oxidation and reduction:

$$4Fe + 3O_2 \rightarrow 2Fe_2O_3$$

Oxidation is a chemical process in which a substance loses electrons. In rusting, iron is changing from its metallic state to its iron(III) (Fe^{3+}) state. We say it is being oxidized:

$$4Fe \rightarrow 4Fe^{3+} + 12e^-$$

The e^- stands for electron. At the same time, oxygen is changing from its molecular state to its charged state:

$$3O_2 + 12e^- \rightarrow 6O^{2-}$$

Oxygen accepts the electrons removed from the iron and is said to be reduced. *Reduction* is a chemical process in which a substance gains electrons. An oxidation cannot take place without a reduction because electrons have to go somewhere; free electrons are not usually found in nature. Some substance must accept the electrons that are lost. Oxidation-reduction reactions are sometimes called *redox reactions*.

Electrons are not easy to remove from covalent compounds unless an entire atom is removed. In living cells oxidation almost always involves the removal of a hydrogen atom from a compound. Reduction often involves a gain in hydrogen atoms. As will be discussed in later chapters, redox reactions are an essential part of photosynthesis, cellular respiration, and other aspects of metabolism.

INORGANIC COMPOUNDS

Chemical compounds can be divided into two broad groups-inorganic and organic. *Inorganic compounds* are relatively small, simple substances. *Organic compounds* are usually large and complex and always contain carbon. Among the biologically important groups of inorganic compounds are water, simple acids and bases, and salts. Organisms depend upon appropriate amounts of these substances for fluid balance, acid-base balance, and many cell activities, such as transporting materials through cell membranes.

Water

Water is an essential ingredient of life. It accounts for about 80% of the weight of an average active cell. In fact, the human body is about 70% water by weight. Many organisms make their homes in lakes, rivers, or the sea, and the cells of terrestrial organisms are bathed in body fluids composed largely of water.

Polar Properties

The physical and chemical properties of water permit life to exist on our planet. Water molecules are *polar*; that is, they bear a partial positive and a partial negative charge. The water molecules in liquid water and in ice are held together in part by hydrogen bonds (Fig. 2-6). The hydrogen atom of one water molecule, with its partial positive charge, is attracted to the oxygen atom of a neighboring water molecule, with its partial negative charge, forming a hydrogen bond. Each water molecule can form hydrogen bonds with a maximum of four neighboring water molecules.

Properties as a Solvent

Because its molecules are polar, water is an excellent **solvent,** a liquid capable of dissolving many polar substances. For example, the ions of a salt, such as sodium chloride, are held together by strong ionic bonds; in fact, they form a crystal lattice. Considerable energy is required to pull the positively and negatively charged ions apart. However, when the sodium chloride is placed in water, the strong electrical attractions between the polar water molecules and Na^+ and Cl^- ions result in the formation of stable dissociated Na^+ and Cl^- ions.

$$\underset{\text{Sodium chloride}}{NaCl} \xrightarrow{\text{in } H_2O} \underset{\text{Sodium ion}}{Na^+} + \underset{\text{Chloride ion}}{Cl^-}$$

Because of its solvent properties and its tendency to cause the ionization of compounds in solution, water is important in facilitating chemical reactions.

Cohesive and Adhesive Forces

Water exhibits both cohesive and adhesive forces. Water molecules have a very strong tendency to stick to each other; that is, they are *cohesive*. This is due to the hydrogen bonds among the molecules. Water molecules also stick to many other kinds of substances (i.e., those substances that have charged groups of atoms or molecules on their surfaces). These *adhesive* forces explain how water makes things wet. Water has a high degree of surface tension because of the cohesiveness of its molecules; its molecules have a much greater attraction for other water molecules than for molecules in the air. Thus water molecules at the surface crowd together, producing a strong layer as they are pulled downward by the attraction of other water molecules beneath them.

Adhesive and cohesive forces account for the tendency, termed *capillary action*, of water to rise in very fine tubes. Capillary action plays some part in the rise of water through the stems of plants to their leaves.

Temperature Stabilization

Other properties are responsible for water's ability to minimize temperature changes. Water has a high specific heat. *Specific heat* is the amount of heat required to raise the temperature of 1 gram of water 1°C. The high specific heat of water is due to its hydrogen bonds. For the temperature of a substance to be raised, heat energy must be added to make its molecules move faster-to increase the kinetic energy of the molecules. Before -stet molecules can move freely, some of the hydrogen bonds bolding the water molecules together must be broken. Thus a great deal of the heat energy added to water is used up in breaking the hydrogen bonds, and only a part of this energy is available to speed the motion of the water molecules (increase the temperature of the water).

Because so much heat input (or heat loss) is required to raise (or lower) the temperature of water, the oceans and other large bodies of water have relatively constant temperatures. Thus the aquatic environment provides the multitude of organisms that inhabit it with a relatively constant environmental temperature. The water within all organisms, even those that live on land, contributes to a relatively constant internal temperature. This is important because metabolic reactions can take place only within a relatively narrow temperature range.

Because its molecules are held together by hydrogen bonds, water also has a high heat of vaporization, another property that helps stabilize temperature. More than 500 calories are required to change 1 gram of liquid water into 1 gram of water vapor. A calorie is a unit of heat energy (defined as 4.184 joules). Because of the heat of vaporization, we can rid ourselves of excess heat by the evaporation of sweat. And plants can remain cool in the midday heat by evaporating water from their surface.

Although most substances become more and more dense as the temperature decreases, water reaches its maximum density at 4°C and then begins to expand again as the temperature decreases. Hydrogen bonds become more rigid and ordered, and ice floats upon the denser cold water. This important property of water explains why lakes and ponds freeze from the surface down rather than from the bottom up. The sheet of ice that forms at the pond surface insulates the water below from the wintry chill so that it is less likely to freeze. Further, organisms that inhabit northern lakes and ponds are able to carry on their life activities despite the frigid winter.

ACIDS AND BASES

An *acid* is a compound that ionizes in solution to yield hydrogen ions (H^+)[1] and an anion. An acid is a proton *donor*. Acids turn blue litmus paper red and have a sour taste. Hydrochloric acid (HCl) and sulfuric acid (H_2SO_4) are examples of inorganic acids. The strength of an acid depends upon the degree to which it ionizes in water. Thus HC1 is a very strong acid because most of its molecules dissociate, producing hydrogen and chloride ions.

$$\underset{\text{Hydrochloric acid}}{HCl} \xrightarrow{H_2O} \underset{\text{Hydrogen ion}}{H^+} + \underset{\text{Chloride ion}}{Cl^-}$$

A base is defined as a proton *acceptor*. Most bases are substances that yield a hydroxide ion (OH^-) and a cation when dissolved in water. Bases turn red litmus paper blue and feel slippery to the touch. Sodium hydroxide (NaOH) and aqueous ammonia (NH_4OH) are inorganic bases.

$$\underset{\text{Sodium hydroxide}}{NaOH} \longrightarrow \underset{\text{Sodium ion}}{Na^+} + \underset{\text{Hydroxide ion}}{OH^-}$$

Since the concentration of hydrogen or hydroxide ions is usually small, it is convenient to express the degree of acidity or alkalinity in a solution in terms of pH, formally defined as the logarithm of the reciprocal of the hydrogen ion concentration, log ($1/[H^+]$). The pH scale is logarithmic, extending from 0, the pH of a strong acid, such as HCl, to 14, the pH of a strong base, such as NaOH. The pH of pure water is 7. Even though water does ionize slightly, the concentrations of H^+ ions and OH^- ions are exactly equal. Solutions with a pH of less than 7 are acidic and contain more H^+ ions than OH^- ions. Solutions with a pH greater than 7 are alkaline, or basic, and contain more OH^- ions than H^+ ions. Because the scale is logarithmic, a solution with a pH of 6 has a hydrogen ion concentration that is ten times greater than a solution with a pH of 7 and is much more acidic. A pH of 5 represents another tenfold increase, and so a solution with a pH of 4 is 10 × 10 or 100 times more acidic than a solution with a pH of 6. The contents of most animal and plant cells are neither strongly acidic nor alkaline but are an essentially neutral mixture of acidic and basic substances. Life is not compatible with any considerable change in the pH of the cell.

SALTS

When an acid and a base are mixed together, the H^+ of the acid unites with the OH^- of the base to form a molecule of water. The remainder of the acid (anion) combines with the remainder of the base (cation) to form a *salt*. This type of reaction is called acid-base neutralization. Hydrochloric acid reacts with sodium hydroxide to form water and sodium chloride:

$$\underset{\text{Hydrogen chloride}}{HCl} + \underset{\text{Sodium hydroxide}}{NaOH} \rightarrow \underset{\text{Water}}{H_2O} + \underset{\text{Sodium chloride}}{NaOH}$$

A *salt* can be defined as a compound in which the hydrogen atom of an acid is replaced by some other cation. A salt contains a cation other than H^+ and an anion other than OH^-. Sodium chloride, NaCl, is a compound in which the hydrogen ion of HCl has been replaced by the cation, Na^+.

When a salt, an acid, or a base is dissolved in water, its constituent ions separate. Because these charged particles can conduct an electric current, these substances are called *electrolytes*. Sugars, alcohols, and many other substances do not ionize when dissolved in water; they do not conduct an electrical current and are termed *nonelectrolytes*.

The cells and extracellular fluids (such as blood) of plants and animals contain a variety of dissolved salts. They are a source of many important mineral ions. Such ions are essential for fluid balance, acid-base balance, nerve and muscle function in animals, blood clotting, bone formation, and many other aspects of body function. Although the concentration of salts in the cells and body fluids of plants and animals is small, these salts are of great importance for normal cell function. The concentrations of the respective cations and anions are kept remarkably constant under normal conditions. Any marked change results in im-paired cellular functions and ultimately in death.

BUFFERS

Many homeostatic mechanisms operate to maintain appropriate pH levels. For example, the pH of human blood is about 7.4 and must be main-tained within very narrow limits. Should the blood become too acidic, coma and death may result. Excessive alkalinity can result in overexcitability of the nervous system and even convulsions. A *buffer* is a substance or combination of substances that resists changes in pH when acids or bases are added. The buffer either accepts or donates hydrogen ions. A buffer consists of a weak acid and a salt of that acid, or a weak base and a salt of the base.

One of the most common buffering systems and one that is impor-tant in human blood is carbonic acid and the bicarbonate ion. Bicarbon-ate ions are formed in the body as follows:

$$\underset{\text{Carbon dioxide}}{CO_2} + \underset{\text{Water}}{H_2O} \rightleftarrows \underset{\text{Carbonic acid}}{H_2CO_3} \rightleftarrows H^+ + \underset{\text{Bicarbonate ion}}{HCO_3^-}$$

As indicated by the arrows, the reactions are reversible. When excess hydrogen ions are present in blood or other body fluids, bicarbonate ions combine with them to form carbonic acid, a weak acid.

$$H^+ + HCO_3^- \rightleftarrows \underset{\text{Carbonic acid}}{H_2CO_3}$$

In this way a strong acid can be converted to a weak acid. The carbonic acid is unstable and quickly breaks down into carbon dioxide and water.

Buffers also work to reduce pH when excessive numbers of hydrox-ide ions are present. A buffer may release hydrogen ions, which combine with the hydroxide ions to form water.

$$OH^- + 2H^+ + CO_3^- \rightleftarrows HCO_3 + H_2O$$

In a way, the chemistry of the carbon atom is the chemistry of life itself. With four electrons in its outer energy level, the carbon atom can share electrons with other atoms, forming four covalent bonds. This ability of carbon atoms yields an immense variety of *organic compounds*, complex compounds that contain carbon. Organic compounds are the main structural components of cells and tissues, the participants in and regulators of thousands of metabolic reactions, and the fuel molecules of living systems.

In all organic compounds the main chain of atoms that makes up the principal axis of the molecule consists of carbon atoms. Thus carbon atoms share electrons with other carbon atoms to form chains of varying lengths. These chains may be unbranched or branched, or the carbon atoms may join to form rings. Adjacent carbon atoms may form single bonds, or by sharing additional pairs of electrons, they may form double (—C=C—) or triple (—C=—C—) bonds.

In this chapter we discuss some of the major groups of organic compounds that are important in living organisms, including carbohydrates, lipids, proteins, and nucleic acids.

CARBOHYDRATES

Familiar to us as sugars and starches, *carbohydrates* serve as fuel molecules and are also important structural components, especially in plant cells. Carbohydrates contain carbon, hydrogen, and oxygen atoms in a ratio of approximately one carbon to two hydrogens to one oxygen $(CH_2O)_n$. The term carbohydrate, meaning "hydrate (water) of carbon," stems from the 2 to 1 ratio of hydrogen to oxygen, the same ratio found in water (H_2O). Carbohydrates may be classified as monosaccharides, disaccharides, or polysaccharides.

Monosaccharides

Monosaccharides are simple sugars that usually contain from three to six carbon atoms. *Glucose* (also called dextrose) and *fructose* are examples of monosaccharides. Each is composed of a single *hexose* unit (consists of six carbon atoms) with the formula $C_6H_{12}O_6$. Recall that such compounds as glucose and fructose, which have identical

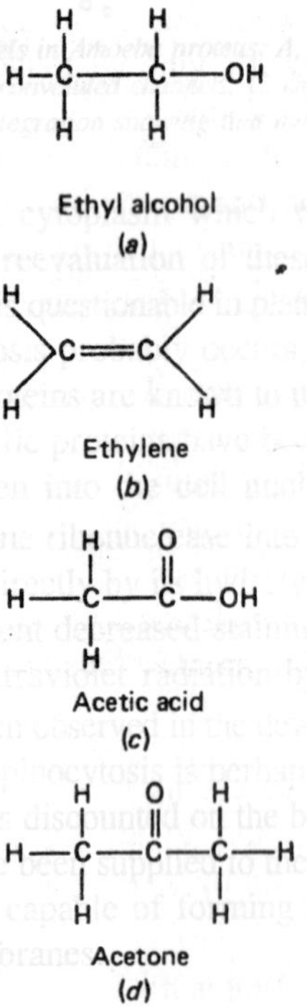

Ethyl alcohol

(*a*)

Ethylene

(*b*)

Acetic acid

(*c*)

Acetone

(*d*)

Figure 2.6 : Some simple organic compounds. Note that each carbon atom has four covalent bonds.

molecular formulas but different arrangements of atoms, are termed *isomers*. Because of their different arrangement of atoms, the two sugars have different chemical properties.

Glucose, often referred to as blood sugar, is the most abundant hexose in the bodies of humans and other animals. Its concentration is kept at a homeostatic level in the blood, and glucose is utilized by the cells as a fuel molecule. The *pentoses* (five-carbon sugars) ribose and deoxyribose are components of nucleotides and nucleic acids.

The "*stick*" formulas in Figure elsewhere in this chapter give a clear but somewhat unrealistic picture of the structures of some common monosaccharides. Actually, molecules are not the simple two-dimensional structures depicted on a printed page. In fact, the properties of each compound depend in part on its three-dimensional structure. Molecules of glucose and other monosaccharides in solution exist mainly as rings.

In some compounds certain atoms can be arranged in more than one position in space around the carbon atom to which they are bonded. Such stereoisomers differ from one another only in some geometric or three-dimensional way. There are two ring forms of glucose, for instance, that differ only in the orientation of an —OH group.

Disaccharides

A *disaccharide* (two sugars) is a carbohydrate that can be degraded into two monosaccharide units. The disaccharide *maltose* (malt sugar) consists of two chemically combined glucose units. **Sucrose,** the sugar we use to sweeten our foods, consists of a glucose unit combined with a fructose unit. *Lactose* (the sugar present in milk) is composed of one molecule of glucose and one of galactose, another hexose monosaccharide. The covalent bond that joins two monosaccharide units is called a *glycosidic bond*.

During digestion maltose is .cleaved (degraded) to form two molecules of glucose:

$$\text{Maltose} + \text{Water} \rightarrow \text{Glucose} + \text{Glucose}$$

Similarly, sucrose is cleaved during digestion to form glucose and fructose:

$$\text{Sucrose} + \text{Water} \rightarrow \text{Glucose} + \text{Fructose}$$

Structural formulas for the compounds in these reactions are shown in Figure elsewhere in this chapter. Because water is added during the cleavage of a disaccharide, this type of reaction is called a *hydrolysis reaction*.

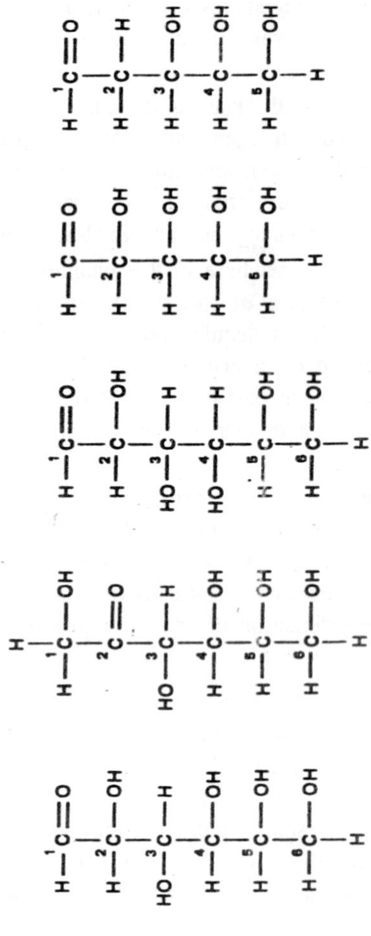

Glucose ($C_6H_{12}O_6$)

Fructose ($C_6H_{12}O_6$)

Galactose ($C_6H_{12}O_6$)

Ribose ($C_5H_{10}O_5$)

Deoxyribose ($C_5H_{10}O_4$)

Figure 2.7 : Structural formulas of some important monosaccharides (simple sugars). The monosaccharides are represented here as straight chains, called stick formulas. Although it is convenient to show monosaccharides in this form, they are more accurately depicted as ring structures. Note that glucose, fructose, and galactose are structural isomers-they have the same chemical formula, $C_6H_{12}O_6$, but their atoms are arranged differently.

Polysaccharides

The most abundant carbohydrates are *polysaccharides*, such as starches, glycogen, or celluloses. A polysaccharide is a single long chain, or a branched chain, consisting of repeating units of a simple sugar, usually glucose. The precise number of sugar units varies, but typically, thousands of units may be present in a single molecule of a polysaccharide. Because they are composed of different stereoisomers of glucose, or because the glucose units are arranged differently, these polysaccharides have very different properties.

Starch is the typical storage form of carbohydrate in plants, whereas *glycogen* (sometimes referred to as animal starch) is the form in which glucose is stored in animal tissues. Glycogen is a highly branched polysaccharide, more water-soluble than plant starch. Glucose cannot be stored as such because its small, readily soluble molecules would leak out of the cells. The larger, less soluble starch and glycogen molecules do not readily pass through the cell membrane. Thus, instead of storing simple sugars, cells store the more complex polysaccharides, such as glycogen, which can readily be broken down into simple sugars.

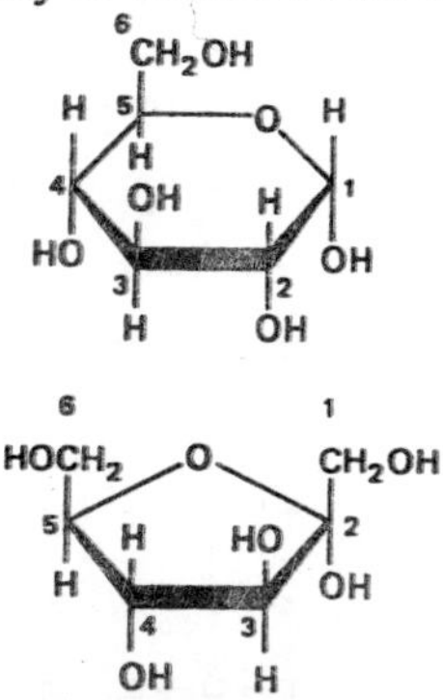

Figure 2.8 : Two monosaccharides, glucose and fructose, drawn to represent their ring structures.

Carbohydrates are the most abundant group of organic compounds on earth, and ***cellulose*** is the most abundant carbohydrate, accounting for 50% or more of all the carbon in plants. Wood is about half cellulose, and cotton is at least 90% cellulose. Plant cells are surrounded by a strong supporting cell wall consisting mainly of cellulose. Cellulose is an insoluble polysaccharide composed of many glucose molecules joined together. The bonds joining these sugar units are different from those in starch and are not split by the enzyme that cleaves the bonds in starch.

The amino sugars glucosamine and galactosamine are modified carbohydrates, compounds in which a hydroxide group (-OH) of a monosaccharide is replaced by an amino group ($-NH_2$; see the section on proteins). Glucosamine is the molecular unit found in chitin; galactosamine is found in cartilage. *Chitin*, a tough modified polysaccharide, is the main component of the external skeletons of insects, crayfish, and other arthropod animals.

LIPIDS

Lipids are a heterogeneous group of compounds that have a greasy or oily consistency and are relatively insoluble in water. Like carbohydrates, lipids are composed of carbon, hydrogen, and oxygen

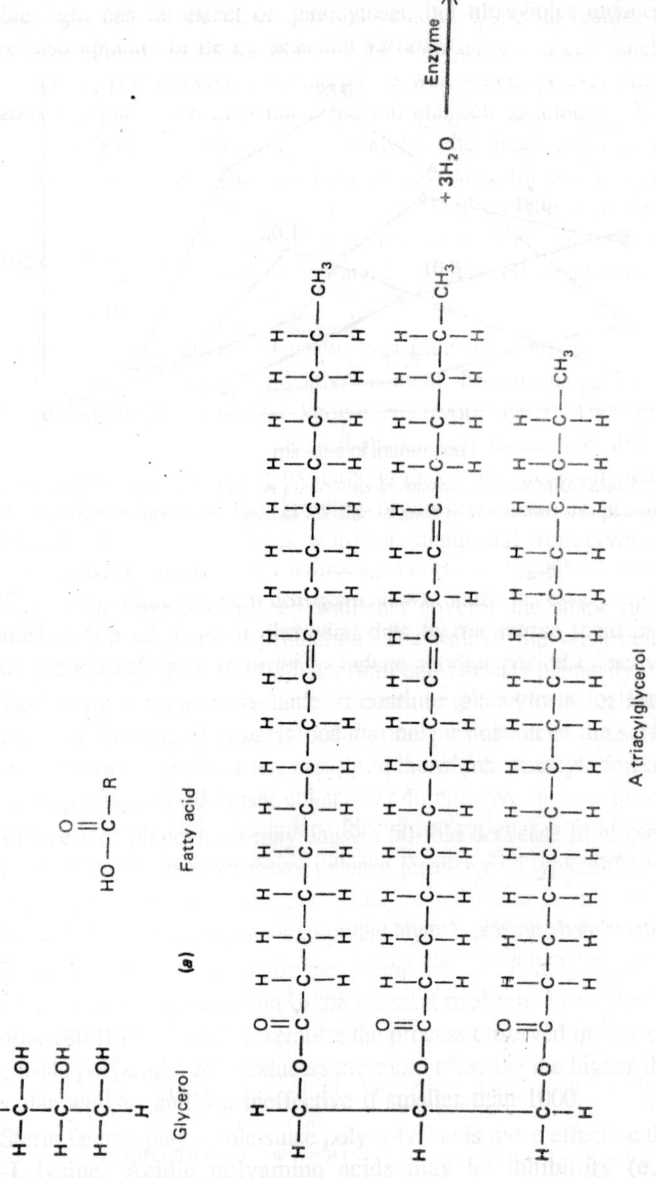

Figure 2.9 : Neutral fats. (a) Structure of glycerol and of a fatty acid. The carboxyl (—COOH) group is present in all fatty acids. The R represents the remainder of the molecule, which varies with each type of fatty acid.

atoms, but they have relatively less oxygen in proportion to the carbon and hydrogen than carbohydrates do. Among the groups of lipids especially significant biologically are the neutral fats, phospholipids, steroids, carotenoids (red and yellow plant pigments), and waxes. Lipids are important biological fuels; they serve as structural components of cell membranes, and some are essential hormones. Glycolipids, lipids that contain watersoluble carbohydrate components, are important in interactions among cells.

Neutral Fats

The most abundant group of lipids in living things is the *neutral fats*. These compounds yield more than twice as much energy per gram as do carbohydrates and are an economical form for the storage of fuel energy. Carbohydrates and proteins can be transformed by enzymes into fats and stored within the cells of adipose tissue.

A neutral fat consists of glycerol joined to one, two, or three molecules of a fatty acid. Glycerol is a three-carbon alcohol' that contains three —OH groups. A fatty acid is a long, straight chain of carbon atoms with a carboxyl group (—COOH) at one end. About 30 varieties of fatty acids are commonly found in animal lipids. They typically have an even number of carbon atoms. For example, butyric acid, present in rancid butter, has four carbon atoms, and oleic acid, the most widely distributed fatty acid in nature, has 18 carbon atoms.

Saturated fatty acids contain the maximum possible hydrogen atoms, while *unsaturated fatty acids* contain carbon atoms that are doubly bonded with one another and are not fully saturated with hydrogens.

Fatty acids with several double bonds are called polyunsaturated fatty acids. Fats containing unsaturated fatty acids are the oils, and most of them are liquid at room temperature. Saturated fats are solids; butter and animal fat are examples. At lease two fatty acids (linoleic and linolenic) are essential nutrients, which must be included in the diet.

When a glycerol molecule combines chemically with one fatty acid, a *monoacylglycerol* (sometimes called monoglyceride) is formed. When two fatty acids combine with a glycerol, a diacylglycerol (or diglyceride) is formed, and when three fatty acids combine with one glycerol molecule, a triacylglycerol (or triglyceride) is formed. In combining with glycerol, the carboxyl end of the fatty acid attaches to one of the —OH groups. This is a dehydration reaction in which H^+ and OH^-, the equivalent of water, is split off. During digestion, on the other hand, the neutral fats may be hydrolyzed to produce fatty acids and glycerol.

Phospholipids

Phospholipids are important constituents of the cell membranes of plants and animals. A phospholipid consists of a glycerol molecule

Phosphatidic acid

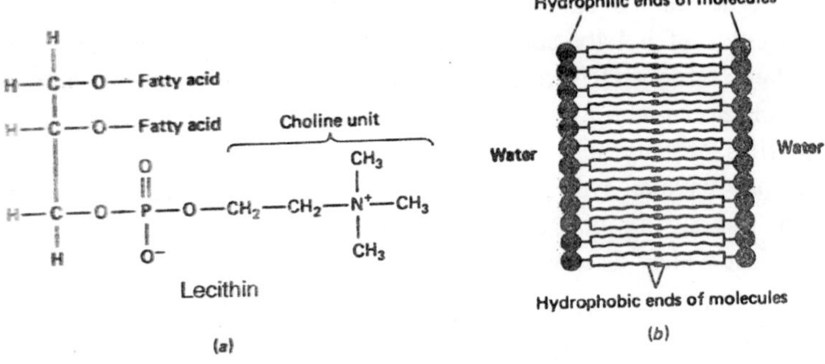

Figure 2.10 : Phospholipids. (a) Many phospholipids are derivatives of phosphatidic acid, a compound consisting of glycerol chemically combined with two fatty acids and a phosphate group. Lecithin is a phospholipid found in cell membranes. It forms when phosphatidic acid combines with the compound choline. (b) A lipid bilayer, such as is found in cell membranes.

attached to one or two fatty acids, but the glycerol is also bonded to phosphorus, which is part of an organic base. Phospholipids also usually contain nitrogen. (Note that phosphorus and nitrogen are absent from neutral fats.)

The two ends of the phospholipid molecule differ physically as well as chemically. The fatty acid portion of the molecule is *hydrophobic* (water-hating) and is not soluble in water. However, the portion composed of glycerol and the organic base is ionized and readily watersoluble. This end of the molecule is said to be *hydrophilic* (waterloving). The polarity of these lipid molecules causes them to take up a certain configuration in the presence of water, with their hydrophilic water-

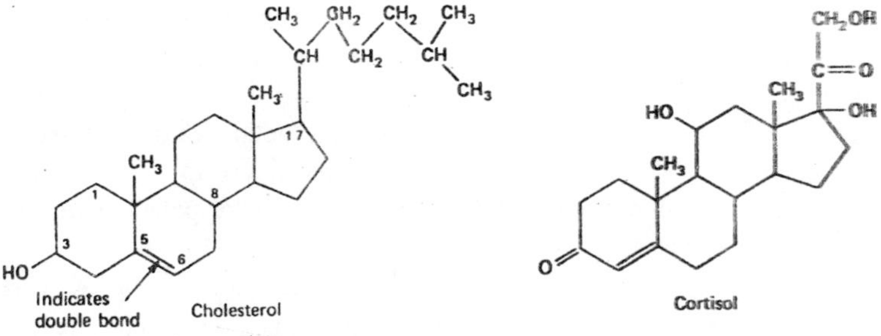

Figure 2.11 : Steroids. All steroids have the basic skeleton of four interlocking rings of carbon atoms. Note that a carbon atom is present at each point in each ring. Each of the first three rings contains six carbon atoms, and the fourth ring contains five. For simplicity, hydrogen atoms have not been drawn within the ring structures.

soluble heads facing outward toward the surrounding water. The hydrophobic tails face in the opposite direction. The cell membrane is a lipid bilayer composed of two layers of phospholipid molecules with their hydrophobic tails meeting in the middle and their hydrophilic heads oriented toward the outside and inside surfaces of the cell membrane.

Steroids

Although steroids are classified as lipids, their structure is quite different from other lipids. A *steroid* molecule contains carbon atoms arranged in four interlocking rings; three of the rings contain six carbon atoms and the fourth contains five. The length and structure of the side chains that extend from these rings distinguish one steroid from another.

Among the steroids of biological importance are cholesterol, bile salts, the male and female sex hormones, and the hormones secreted by the adrenal cortex. Cholesterol is a structural component of animal cell membranes. Steroid hormones regulate certain aspects of metabolism in a variety of animals, including vertebrates, insects, and crabs.

PROTEINS

Proteins serve as important structural components of cells and tissues, so growth and repair, as well as maintenance of the organism, depend upon an adequate supply of these compounds. Some proteins serve as *enzymes*, catalysts that regulate the thousands of different chemical reactions that take place in a living system (see Focus on Classifying Proteins).

The protein constituents of a cell are the clue to its life-style.

Each cell type has characteristic types, distributions, and amounts of protein that determine what the cell looks like and how it functions. A muscle cell is different from other cell types by virtue of its large content of the contractile proteins myosin and actin, which are largely responsible for its appearance as well as its ability to contract. The protein hemoglobin, found in red blood cells, is responsible for the specialized function of oxygen transport.

Most proteins are species-specific; that is, they vary slightly in each species, so the protein complement (as determined by the instructions in the genes) is also mainly responsible for differences among species. Thus the types and distributions of proteins in the cells of a dog vary somewhat from those in the cells of a fox or a coyote. The degree of difference in the proteins of two species is thought to depend upon evolutionary relationships. Organisms less closely related by evolution have proteins that differ more markedly than those of closely related forms. Some proteins differ slightly even among individuals of the same species, so each individual is biochemically unique. Only genetically identical organisms-identical twins or members of closely inbred strains of organisms-have identical proteins.

Amino Acid Structure

A basic knowledge of protein chemistry is essential for understanding nutrition as well as other aspects of metabolism. Proteins are composed of carbon, hydrogen, oxygen, nitrogen, and usually sulfur. Atoms of these elements are arranged into molecular subunits called *amino acids*. All of the more than 20 kinds of amino acids commonly found in proteins contain an amino group ($-NH_2$) and a carboxyl group (—COOH) bonded to the same carbon atom, but they differ in their side chains, abbreviated as "R" groups. *Glycine*, the simplest amino acid, has a hydrogen atom as its R group or side chain; alanine has a methyl ($—CH_3$) group.

Amino acids combine chemically with one another by bonding the carboxyl carbon of one molecule to the amino nitrogen of another. This covalent bond linking two amino acids is referred to as a *peptide bond*. When two amino acids combine, a *dipeptide* is formed; a longer chain of amino acids is a *polypeptide*.

Each protein may contain hundreds of amino acids joined in a specific linear order. An almost infinite variety of protein molecules is possible. It should be clear that the various proteins differ from one another in the number, types, and arrangement of amino acids

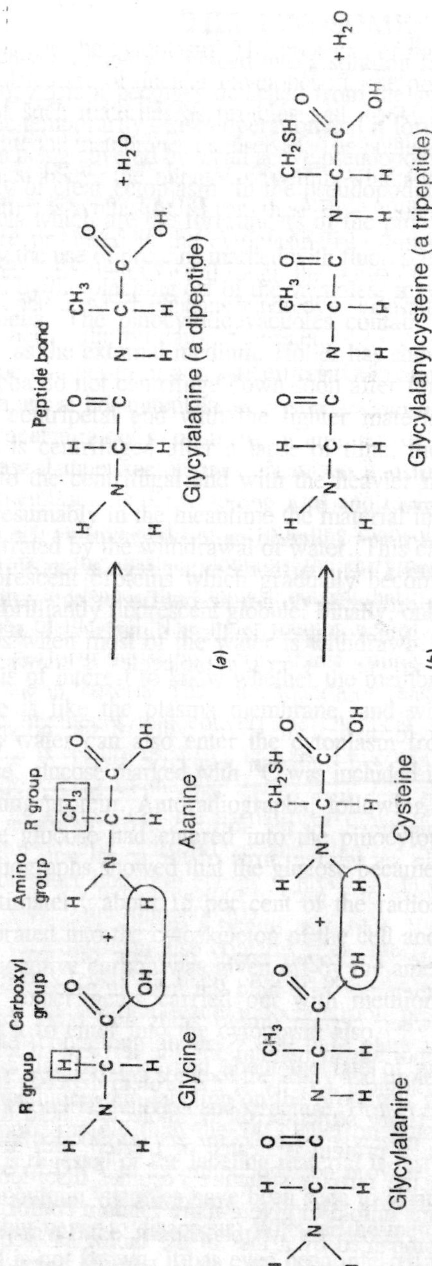

Figure 2.12 : (a) Formation of a dipeptide. Two amino acids combine chemically to form a dipeptide. Water is produced as a by-product during this reaction. (b) A third amino acid is added to the dipeptide to form a chain of three amino acids (a tripeptide or small polypeptide). The bond between two amino acids is a peptide bond.

that they contain. The 20 types of amino acids found in biologic proteins may be thought of as letters of a protein alphabet. Each protein is a word made up of amino acid letters.

With some exceptions, plants can synthesize all their needed amino acids from simpler substances. The cells of humans and animals in general can manufacture some, but not all, of the various kinds of biologically significant amino acids if the proper raw materials are available. Those that animals cannot synthesize but must obtain in the diet are known as *essential amino acids*. Animals differ in their biosynthetic capacities; what is an essential amino acid for one species may not be essential in the diet of another.

Protein Structure-Levels of Organization

Several levels of organization can be distinguished in the protein molecule. The sequence of amino acids in a polypeptide chain constitutes its *primary structure*. This sequence, as we shall see later, is specified by the instructions in a gene.

The *secondary structure* of protein molecules involves the coiling of the peptide chain into a helix or some other regular conformation (shape). Peptide chains ordinarily do not lie out flat or coil randomly, but undergo coiling to yield a specific three-dimensional structure. A common secondary structure in protein molecules is known as the α-helix.

This involves the formation of spiral coils of the polypeptide chain. The α-helix is a very uniform geometric structure with 3.6 amino acids occupying each turn of the helix. The helical structure is determined and maintained by the formation of hydrogen bonds between amino acids in successive turns of the spiral coil.

The *tertiary structure* of a protein molecule involves the folding of the a-helix upon itself, folds that impart a specific overall structure to the protein molecule. Hydrogen bonds between one part of the peptide chain and another part help hold the folds in place. Disulfide bonds (—S—S—) between certain amino acids and other covalent bonds may also be important in maintaining the tertiary structure of many proteins. The biological activity of the protein depends in large part on the specific tertiary structure of the molecule, held together by these bonds. When a protein is heated or treated with any of a number of chemicals, the tertiary structure becomes disordered and the coiled peptide chains unfold to give a more random conformation. This unfolding is accompanied by a loss of the biological activity of the protein, for example, its ability to act as an enzyme. This change

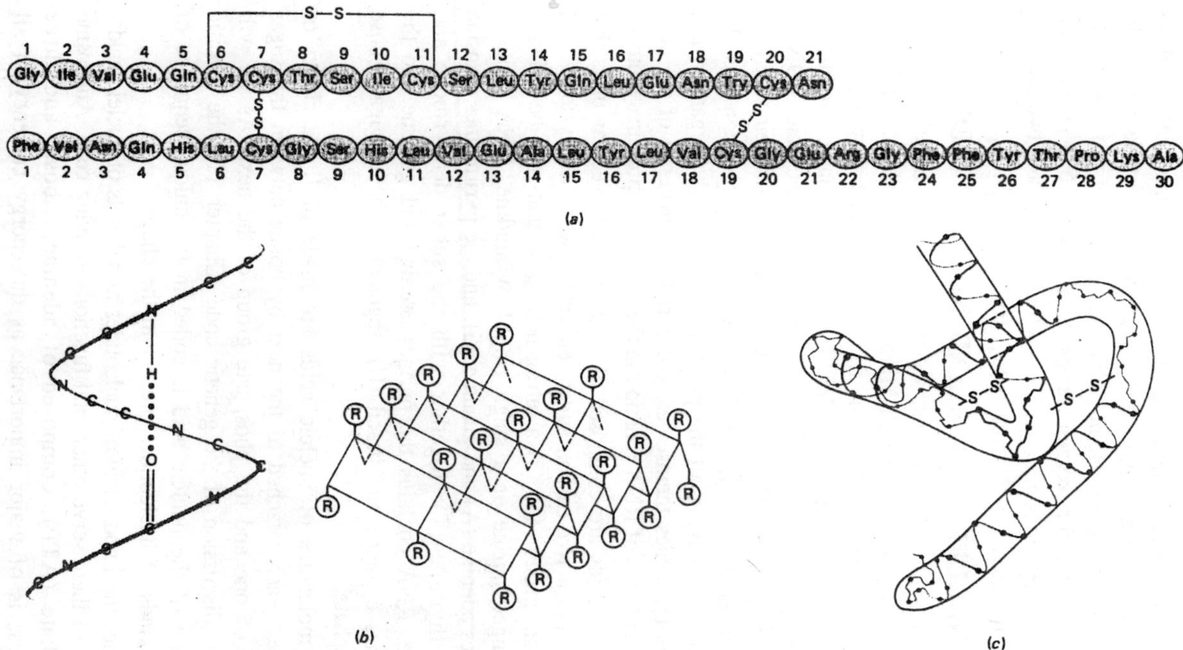

Figure 2.13 : Protein structure. (a) The primary structure of the two polypeptide chains that make up the protein insulin. (b) The secondary structure of proteins is commonly an alpha helix. (c) The tertiary structure results from the coiling and folding of the alpha helix (or other secondary structure) into an overall globular or other shape.

is termed *denaturation* of the protein. Proteins composed of two or more subunits have a *quaternary structure*. This refers to the combination of two or more like or unlike peptide chain subunits, each with its own primary, secondary, and tertiary structures, to form the biologically active protein molecule. Hemoglobin, the protein in red blood cells that is responsible for oxygen transport, consists of 574 amino acids arranged in four polypeptide chains—two identical α and two identical β chains. Its chemical formula is $C_{3032}H_{4816}O_{872}N_{780}S_8Fe_4$.

NUCLEIC ACIDS AND NUCLEOTIDES

Nucleic acids, like proteins, are large, complex molecules. They were first isolated by Friedrich Miescher in 1870 from the nuclei of pus cells, and their name stems from the fact that they are acidic and were first identified in nuclei. There are two classes of nucleic acids, ribonucleic acids (RNA) and deoxyribonucleic acids (DNA); the different kinds of RNA and DNA vary in some of their structural components and in their metabolic functions. DNA contains the instructions for making all the proteins needed by the organism and constitutes the genes themselves, the hereditary material of the cell. The various kinds of RNA function in the process of protein synthesis.

Nucleic acids are composed of *nucleotides*, molecular units that consist of (1) a five-carbon sugar, either ribose or deoxyribose, (2) a phosphate group, and (3) a nitrogenous base that may be either a double-ringed purine or a single-ringed pyrimidine. DNA contains the purines adenine (A) and guanine (G), and the pyrimidines cytosine (C) and thymine (T), together with the sugar deoxyribose and phosphate. RNA contains the purines adenine and guanine and the pyrimidines cytosine and uracil (U), together with the sugar ribose and phosphate.

The molecules of nucleic acids are made of linear chains of nucleotides, each attached to the next by bonds between the sugar molecule of one and the phosphate group of the next. As we will see in our discussion of the genetic code (Chapter 11), the specific information of the nucleic acid is coded in the unique sequence of the four kinds of nucleotides present in the chain.

Besides the importance of nucleotides as subunits of nucleic acids, a number of them serve other vital functions in living cells. Adenosine triphosphate (ATP), composed of adenine, ribose, and three phosphates, is of major importance as the energy currency of all cells.

The two terminal phosphate groups are joined to the nucleotide by special "energy-rich" bonds, indicated by the ~ P symbol. These are energyrich bonds in the sense that much free energy is released when the bonds are hydrolyzed. The biologically useful energy of these bonds can be transferred to other molecules. Most of the chemical energy of the cell is stored in these high energy phosphate bonds of ATP, ready to be released when the phosphate group is transferred to another molecule.

A nucleotide may be converted by enzymes called cyclases to a cyclic form. ATP, for example, is converted to cyclic adenosine monophosphate (cyclic AMP) by the enzyme adenylate cyclase. Cyclic nucleotides play important roles in mediating the effects of hormones and in regulating various aspects of cellular function.

Cells contain several dinucleotides that are of great importance in metabolic processes. For example, nicotinamide adenine dinucleotide (NAD^+) is very important as a primary electron and hydrogen acceptor and donor in biological oxidations and reductions within cells.

3

STRUCTURE OF CELL

All organisms from the simplest to the most complex plant or animal are composed of cells. Some organisms, like a bacterium or an amoeba, consist of a single cell. A complex organism like an adult human is made up of more than one hundred trillion (10^{14}) cells. The appearance, behavior, activities, and functions of all organisms represent the sum total of the properties of the cell or cells of which they are made. Therefore, the study of the cell is overwhelmingly important for understanding life in all its forms.

The contemporary study of the cell is an enormous enterprise that combines many scientific disciplines and methods, including microscopy, genetics, biochemistry, biophysics, and physiology. New information has accumulated rapidly, particularly in the last 30 years, and a great deal about cells is now understood-how they function, how they are constructed, and how they work together within a multicellular organism. However, much remains to be learned.

The beginnings of cell biology go back more than three centuries to disveries by the first pioneering microscopists. In 1665 the Englishman Robert Hooke observed the cell walls of cork with a primitive microscope and also discovered the cellular nature of a living plant leaf. A leaf was easily studied with Hooke's microscope because a leaf is thin enough for light to pass through, allowing its large cells to be seen. Hooke applied the term *cell* to describe the microscopic units he discovered. A contemporary of Hooke, Antome van Leeuwenhoek, discovered sperm cells, red blood cells, and many kinds of microorganisms using a microscope consisting of a single lens. These early observations set the stage for almost two centuries of microscopic studies. As microscopes gradually improved, the cell nucleus was discovered, and the cellular nature of all plants and animals came to be recognized. By the 1830s enough observations had been accumulated to allow Schleiden, a botanist, and Schwann, a zoologist, to propose that all organisms are constructed of cells, a statement now known as the *cell doctrine*. This pronouncement constitutes the first major tenet upon which the contemporary science of cell biology is founded. Following the formulation of the cell doctrine, biologists established that new cells do not arise *de novo* but come only by *cell division*, that is, division of a preexisting cell into two daughter cells. By the end of the nineteenth century, cell biologists had discovered chromosomes and described *mitosis*—the distribution a at cell division of chromosomes to daughter cells. Studies that followed soon led cell biologists to conclude that chromosomes are the hereditary material, and that during cell division, the process of mitosis distributes this material equally to the two daughter cells.

Although microscopic studies from 1665 to 1900 convinced cell biologists that the cell is the building block of all multicellular organisms, a major question remained: Were microorganisms such as bacteria and protozoa single-celled organisms or were they instead noncellular or acellular? As biologists came to understand the genetic, physiological, and biochemical properties of cells, it became clear that microorganisms are single, free-living cells. All modern research recognizes that in both unicellular and multicellular organisms the cell is the fundamental structural unit, housing the genetic material and the biochemical organization that account for the existence of life.

This book describes the main elements of what has been learned so far about cells: how they are constructed and how they carry out such activities as selfmaintenance, growth, cell division, differentiation, movement, sensory perception, and cell-cell communication. The discussions of cell activities and structures are based on integration of various kinds of information derived both from the more traditional tools (microscopy, physiology, and conventional genetics) and from the more recent approaches of biochemistry, macromolecular chemistry, and molecular genetics. Integration of information derived from these diverse disciplines should eventually provide a complete understanding of all the functional and structural properties of cells.

ALL CELLS POSSESS THE SAME BASIC PROPERTIES

We speak of the cell, yet one must be aware of the existence of millions of different species of cells that show tremendous diversity in structure and metabolic capabilities. A bacterium, a yeast, an amoeba, a plant cell, and a liver cell would appear to be so different in structure and life-style that they might seem to have little in common, yet the similarities among these diverse cell types are more profound than the differences. The main properties in common are the following:

1. All cells store information in genes made of DNA.
2. The genetic code used in the genes of all cells is, with minor exceptions, the same.
3. All cell types decode the genes in their DNA by an RNA system that translates genetic information into proteins.
4. All cells synthesize proteins using a structure called the ribosome.
5. Proteins govern function and structure in all cells.
6. All cells need energy to maintain their internal environment and to drive the synthesis of their complex constituents. All cells use the molecule ATP as the currency for transfer of energy from energy sources to energy needs.
7. All cells are enclosed by a plasma membrane composed of proteins and a double layer of lipid molecules.

Because of these similarities, it is possible to generalize about the structures and functions of cells. One can begin constructing a set of principles of function and structure that has been present throughout the evolution of the great diversity of cell types that now exists.

CLASSIFICATION OF CELLS-PROKARYOTES AND EUKARYOTES

On the basis of certain major properties, all cell types currently found on this planet fall into two major groups: the *prokaryotes* and the *eukaryotes*. *Karyote* is a word root meaning nucleus. The word *prokaryote* (before a nucleus) therefore designates cells that do not have a structurally delineated unit containing the genetic material (a nucleus). *Eu* means true; hence, the word *eukaryote* (true nucleus) designates cells that have a well defined nucleus whose chromosomal material is separated from the remainder of the cell contents (the cytoplasm) by a nuclear envelope. Prokaryotes do not have a nuclear envelope. The nucleus of a prokaryote is less well delineated and is usually called a *nucleoid*. The prokaryotic cells are the bacteria, and all are unicellular. These include the blue-green bacteria, or *cyanobacteria,* which were until recently called blue-green algae. All other cell types are eukaryotes. These include many unicellular organisms, for example, protozoa, fungi, and some algae, as well as all multicellular plants and animals.

Originally, prokaryotes were distinguished from eukaryotes on the basis of a single characteristic, the absence of a nuclear envelope. Subsequently, other equally distinct differences have been discovered. The main differences are listed in Table elsewhere in this chapter. Some of the characteristics listed in Table elsewhere in this chapter need explanations, for example introns in genes; the explanations are given in later chapters. Prokaryotic cells are both smaller and structurally less complex than eukaryotic cells. In addition, eukaryotes contain far more DNA than prokaryotes. Among the hundreds of eukaryotes examined, those with the least amount of DNA (yeast cells) still have several times more than the prokaryote thought to possess the most DNA (the bacterium *Serratia marcescens*). This difference in DNA content reflects the fact that eukaryotes are genetically more complex than prokaryotes.

It is thought that prokaryote-like cells preceded eukaryotic cells in evolution and that the first eukaryote evolved from some type of prokaryote, probably a billion or more years ago. We have no clear idea which difference between prokaryotes and eukaryotes was the first to arise; hence, we do not know which of the differences was fundamental in the origination of eukaryotes. Presumably, a prokaryotic cell acquired one or another of the key characteristics for eukaryotes listed in Table elsewhere in this chapter, and this created the potential for subsequent evolution of the other distinguishing characteristics of eukaryotes. For instance, the acquisition through evolution of an unusually large amount of DNA could have been the innovative step that first separated an emerging eukaryote from prokaryotes. What could have brought about an increase in DNA content, however, is not known. Whatever the first step in the origination of eukaryotic cells, that step must have created the potential for great evolutionary diversification. The range in function and structure among contemporary cells is many times greater among eukaryotes than it is among prokaryotes.

Throughout this book comparisons between prokaryotes and eukaryotes enter into discussions of cell function and structure. Such comparisons often yield insight into cell organization, operation, and evolution. The following section briefly characterizes prokaryotic cells. The general properties of eukaryotic cells are then outlined through descriptions of a yeast cell, an amoeba, a plant cell, and several mammalian cells.

PROKARYOTIC CELLS

Prokaryotes are divided into two major groups, the *eubacteria* and the *archaebacteria*. Both groups share many properties-those listed in Table elsewhere in this chapter—that clearly separate them from eukaryotes. Most species of bacteria are eubacteria. Eubacteria can be divided into two major groups: the *nonphotosynthetic* and *photosynthetic bacteria*. These two groups are discussed in the next two sections. The bacterium *Escherichia coli* is used as an example of nonphotosynthetic eubacteria.

Nonphotosynthetic Eubacteria

Escherichia coli is a bacterium that commonly inhabits the intestinal tract of humans and other animals. It is a cylindrical cell about 2 μm long and 1 μm in diameter, with a volume of about 1.6 μm. On its surface are a number of filamentous appendages called *flagella*, usually six, by which it rapidly propels itself. One cm^3 (about one gram) of packed *E. coli* contains about 50×10^9 cells. An individual cell grows by increasing its length while maintaining a constant diameter. The cell divides into two daughters by the forming of a partition through the middle of the cylinder.

The genetic constitution of *E. coli* allows the organism to grow and divide in a medium containing only a few kinds of inorganic ions and a source of organic carbon, for example, the sugar glucose. Thus, the DNA of this bacterial cell contains genes for all of the enzymes needed for the synthesis of all the amino acids, nucleosides, fatty acids, and other components needed to make macromolecules, using only a simple organic molecule, such as glucose, and inorganic salts as the starting materials.

In glucose-containing medium, *E. coli* doubles in size and divides every 40 minutes at the optimum temperature of 37°C. If amino acids, nucleosides, and other nutritionally useful organic molecules are added to the minimal culture medium, the rate of cell growth and division increases. This increase occurs because provision of useful nutrients relieves the cell of the need to synthesize those components. By addition of a rich variety of nutrients to the medium, an *E. coli* cell approaches the upper limit in its reproductive rate, doubling all of its contents and dividing every 20 minutes. The upper limit to growth rate is probably set by the maximum rate at which *E. coli* can synthesize macromolecules. With a generation time of only 20 minutes, a single cell could give rise to 2^{36} cells (more than 3×10^{11} cells) in just 12 hours.

Prokaryotic cells usually grow and reproduce much more rapidly than eukaryotes. For instance, for a mammalian cell, seven hours is about the shortest *generation time* or *cell cycle time*—the time a cell

Table 3.1 : Major Differences Between Prokaryotic and Eukaryotic Cells

Prokaryotic	Eukaryotic
1. No nuclear envelope	Nuclear envelope present
2. No nucleolus	One or more nucleoli
3. No histone proteins associated with DNA	Histories bound to DNA
4. DNA content ranges from 750,000 base pairs to 5 x 10^5 base pairs	DNA content ranges from about 1.5×10^7 to 1.5×10^{11} base pairs
5. Genes generally lack introns	Most genes contain introns
6. One chromosome	Two or more chromosomes
7. Small cell size; usually no more than several μm^3 in volume	Large cell size; usually from several μm^3 to several mm^3 in volume
8. Lack well developed intracellular membranous organelles except for photosynthetic membranes in some bacteria	Contain extensive membr--ane systems and membran--ous organelles such as mitochondria, chloroplasts, Golgi complex, and endop-lasmic rcticulum
9. Lack microtuhules, microfilaments, and intermediate filaments	Contain inicrotubules, microfilaments, and interm-ediate filaments
10. Sterols usually not present in the plasma membrane	Sterols present in the plasma membrane

needs to go from one cell division to the next. Some protozoa, which are unicellular eukaryotes, have generation times as short as two hours in a nutritionally rich medium, and one kind of yeast cell can divide every 75 minutes; but none of the eukaryotes approaches the rapid proliferation rates common among prokaryotes.

Prokaryotes are structurally simple cells. Most types have a rigid *cell wall* made of polysaccharides, peptides, and lipids laid down outside the cell. A few types, for example, the small bacteria known as *Mycoplasmas,* lack an extracellular wall. The rigid wall of rod-shaped bacteria maintains the cylindrical shape of the cell. Other species of bacteria have walls that produce a spherical or a spiral form. The cell wall provides mechanical protection, particularly against osmotic pressure.

Immediately inside the cell wall of *E. coli* is the *plasma membrane*, which completely encloses the cell. The plasma membrane consists of a double layer of lipid molecules with many associated protein molecules. In contrast to the cell wall, which is porous and therefore penetrable by molecules and ions, the plasma membrane severely restricts the diffusion of molecules and ions in and out of the cell. Thus, the membrane serves the critical role of retaining desired substances inside the cell, although it also limits diffusion into the cell of environmental substances necessary to sustain cell metabolism. Certain specialized proteins bound to the lipid bilayer of the plasma membrane greatly enhance the inward passage of inorganic ions, sugars, amino acids, nucleosides, and other dissolved materials that are useful to the cell. Other proteins bound in the plasma membrane of a bacterial cell catalyze the process by which the energy contained in organic molecules is converted into a chemically usable form.

The intracellular contents of prokaryotes such as *E. coli* are present in two major structural parts, a nucleoid and the cytoplasm. The nucleoid consists of a single DNA molecule (the chromosome) condensed into an irregularly shaped, fibrous network, which occupies a few percent of the total cell volume. It is thought that the nucleoid is attached at one point to the plasma membrane. This attachment of the chromosome to the membrane may help both in the control of chromosome replication and in the separation of daughter chromosomes during cell division.

The cytoplasm of *E. coli* contains approximately 25,000 tiny particles called *ribosomes*, floating in a solution called the *cytosol*. Each ribosome is a machine for synthesizing proteins. The cytosol, which contains a large variety of ions, small organic molecules, and enzymes, is where the cell carries out most of its metabolic activities.

Much has been learned about the molecular biology of the cell from the study of prokaryotes, in particular from the study of *E. coli*. In part, bacteria were chosen as research materials because they are functionally and structurally far less complex than any of the eukaryotic cells. In addition, the fast growth rate and low number of nutritional requirements of bacteria such as *E. coli* constitute a great practical advantage for research because large numbers of cells can be obtained in a few hours with a simple, inexpensive culture medium.

E. coli is not the smallest type of cell known. Some bacteria, the mycoplasmas have volumes as small as 0.02 μm^3, compared to a minimum volume for *E. coli* of 1.6 μm^3. Mycoplasmas lack cell walls, and their chromosomes can be as small as one-fifth of the chromosome in *E. coli.* These are the smallest chromosomes known among bacteria. The mycoplasmas were identified about 1900 as the cause of respiratory disease in animals and gained attention during World War II as the causative agents of pneumonia among U.S. Army recruits. Mycoplasmas are sometimes referred to as PPLO, which stands for pleuropneumonia-like organisms. Given their small size and small amount of DNA, the mycoplasmas are no doubt genetically and functionally less complex than E. *coli.* However, they require a nutritionally complicated medium for growth and grow slowly and hence are less convenient to use in research. Nevertheless, the study of mycoplasmas has intensified during recent years, and these simplest of known cells may well provide unique insight into principles of cell organization and operation. Mycoplasmas are sometimes referred to as the ***minimum*** cell because they approach the minimum genetic and molecular complexity necessary to sustain the life and reproduction of a cell.

Photosynthetic Bacteria

The photosynthetic bacteria probably arose from nonphotosynthetic bacteria very early in the course of evolution, perhaps as early as 3.1 billion years ago. Most photosynthetic bacteria are ***obligate photoautotrophs***. Photoautotroph means requiring only light, water, inorganic ions, and CO_2; obligate means that for growth light is necessary because these bacteria cannot use organic compounds like sugars as an alternative source of energy. Photosynthetic bacteria are widely distributed in fresh and salt water and in soil.

The enormous mass of photosynthetic bacteria growing in the oceans generates much of the oxygen in the Earth's atmosphere.

An electron micrograph of a section through a photosynthetic bacterium. The cell is enclosed by a rigid wall and, immediately inside the wall, by a plasma membrane. As in other bacteria, the cytoplasm is rich in ribosomes, and a nucleoid is present. In contrast to other kinds of bacteria, however, photosynthetic bacteria often have extensive internal membranes that contain light-absorbing pigments and the machinery for photosynthesis. Photosynthesis is the process by which the energy of light is captured and used to synthesize sugar, starting with carbon dioxide and water.

EUKARYOTIC CELLS

All eukaryotic cells share certain basic properties, but nonetheless the eukaryotes are an extremely diverse group. Four cell types-a yeast cell, an amoeba (both unicellular organisms), a plant cell, and an animal cell-are described here to present an overview of the general features of eukaryotes. However these cells only begin to indicate the full range of diversity four, in the group. Many kinds of eukaryotic cells are used in research, because different cells have special feature that make them suitable for experiments on particuli aspects of cell function and structure. The amoeba, for example, is well suited for experiments requiring tranplantation of a nucleus from one cell to another. Yeast cells are particularly favorable for study of the genet basis of cell operations. Mammalian cells in culture are favorable for studies of cell growth and division. Plant cells are important for the analysis of photosynthesis. Experiments with these and many other cell types will eventually lead to a detailed understanding of the cell in general.

The Yeast Cells

Yeast is a unicellular eukaryote classified as a fungus. In cell size, structural complexity, and DNA content the various species of yeasts are among the simplest known eukaryotic cells. The yeast, *Saccharomyces cerevisiae, is* an oval-shaped cell measuring about 6 μm in its maximum diameter. Its name comes from *saccharo,* which means sugar-a food that the growing organism consumes in large quantities-from *myces,* the Greek word for fungus-and from *cerevisia,* the Latin word for beer.

Yeast cells have a rigid cell wall that governs the shape of the cell and gives it mechanical protection. The wall of the yeast cell is much thicker than the wall of a bacterium and consists primarily of a material (polymerized sugars) that is chemically related to the cellulose of plant cell walls. The wall is outside the plasma membrane. The plasma membrane regulates the passage of ions and molecules in and out of the cell. Unlike bacteria, eukaryotes do not have in their plasma membranes the enzymatic machinery for converting energy to a usable form; in eukaryotes the comparable function is carried out in cytoplasmic organelles called *mitochondria*.

The yeast cell has two major compartments, the nucleus and the cytoplasm. The nucleus contains the *chromosomes*, which appear in the electron microscope as a tangled array of finely dispersed fibers, and which are separated from the *cytoplasm* by the nuclear envelope. The 17 chromosomes in a yeast cell have an aggregate DNA content of 1.5×10^7 base pairs. Hence, this simple eukaryote contains about four times more DNA than does *E. coli*.

The nuclear envelope of the yeast cell and all other eukaryotes is a *double membrane*, whose function is not well understood. It almost certainly helps to regulate the movement of materials between the nucleus and the cytoplasm. The presence of *pores* in the envelope is a general feature of nuclear envelopes. These pores allow the entry and exit of such materials as proteins and RNA, which are too large to penetrate the membranes themselves. The nucleus also contains a roughly spherical body, the *nucleolus*, within which the RNA component of ribosomes is synthesized. From these RNA molecules, and from proteins that are produced in the cytoplasm, ribosomes are assembled in the nucleolus and then move through the pores of the nuclear envelope to the cytoplasm, which contains hundreds of thousands of ribosomes working to synthesize the proteins of the cell.

The cytoplasm of a yeast cell contains a number of mitochondria. Mitochondria are cytoplasmic organelles (an organelle is a structural component in a cell) consisting of an outer delimiting membrane and an array of inner membranes organized into folds called *cristae*. The outer membrane is permeable to ions and molecules moving between the mitochondrion and the surrounding cytosol. The system of inner membranes, the cristae, contains the enzymes that oxidize organic molecules to provide the chemical energy that fuels the many metabolic activities of the cell. The inner membranes of mitochondria, in other words, perform the same energy-transforming function as the plasma membrane in bacteria. The mitochondria of yeasts and other eukaryotes are the organelles that convert the chemical energy in food into a form that is readily used by the cells. In nonphotosynthetic cells, they form 95 percent of the fuel for cellular activities. The remaining 5 percent is produced by the partial breakdown of sugars in the cytosol. In photosynthetic cells (algae and plants), the situation is a little different, since chemical energy is produced in chloroplasts as well as in the cytosol and mitochondria.

A mitochondrion also contains ribosomes, which are used in the synthesis of a few of the many proteins needed to carry out mitochondrial activities. In addition, each mitochondrion possesses a small DNA molecule (containing about 75,000 base pairs in yeast) that contains genes for synthesizing some of the RNA and protein components needed for mitochondrial function and structure. However, most of the proteins of the mitochondrion are encoded by genes in the cell nucleus. The presence of DNA in the mitochondrion provides this organelle with a limited amount of autonomy, although growth, multiplication, and metabolism of the mitochondrion are closely coordinated with the metabolism of the rest of the cell through mitochondrial-cytoplasmic-nuclear interactions.

The cytoplasm of yeast cells contains several other important functional components in addition to ribosomes and mitochondria. Most of these other components are discussed later in this chapter in connection with other kinds of cells in which they are more readily identified by electron microscopy.

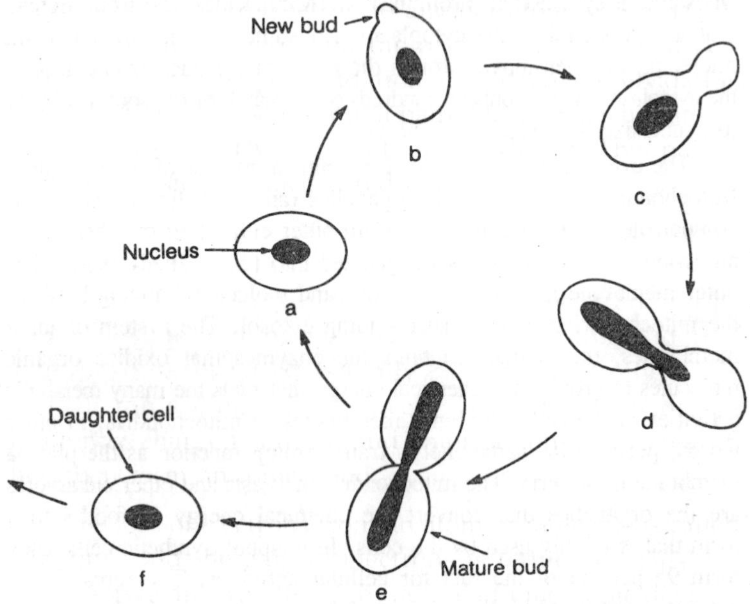

Figure 3.1 : The yeast Saccharomyces cerevisiae reproduced by budding.

In yeast cells that are no longer growing, one or two large *lysosomes* are prominent. Lysosomes are vesicles containing high concentrations of enzymes that break down proteins, polysaccharides, lipids, and nucleic acids. As long as these enzymes remain packaged in lysosomal vacuoles, the cell is protected from the digestive action of its own digestive enzymes. If the cell dies, the lysosomal enzymes are released into the cytoplasm, where they cause rapid destruction of cellular components. This process of selfdigestion is known as *autolysis*. The cytoplasm of yeast cells contains *lipid droplets*, which are storage depots of lipid molecules that can be broken down to yield energy.

In summary, the most simply structured of eukaryotes, exemplified by the yeast cell, has much greater structural complexity than bacterial cells in the form of more DNA, multiple chromosomes, a nucleolus, a nuclear envelope, and a variety of cytoplasmic organelles.

Most yeast species grow in simple media, requiring no more than inorganic ions, a carbohydrate such as glucose, and a few vitamins. In such a medium yeast cells double in number every few hours. Some species reproduce by *fission*, a process whereby a cell splits into two. Budding yeasts like *Saccharomyces* reproduce by forming a bud, which is an unusual mode of cell reproduction among eukaryotes (most divide by fission). The process begins with the formation of a small bud at the cell surface. The bud grows into an incipient daughter cell, and subsequently the nucleus in the parent cell divides by mitosis; the daughter cell receives one of the daughter nuclei while the parent cell retains the other. After mitosis the bridge connecting the mother and daughter cells is closed off and the cells separate from each other.

It might seem more logical for cell biologists to have begun the study of the eukaryotic cell by a thorough analysis of the properties and behavior of one of the simple cell types, such as yeast. This was not done partly because the relative simplicity and experimental advantages of yeasts were not widely recognized until recent years. Furthermore, the importance of studying the simplest available eukaryote was perhaps not adequately appreciated. Also, in other kinds of eukaryotic cells some of the cytoplasmic organelles are more abundant and therefore more immediately amenable to experimental manipulation. Now that molecular biology has begun to reveal a uniformity in the basic molecular organization and operation of all eukaryotes, the logic of intense study on yeasts has become more compelling. Research on this cell type, particularly on the genetic basis of cell properties, is rapidly expanding.

Amoeba

The large, freshwater amoebae present a sharp contrast to the yeast cell. Amoebae are not only among the largest of eukaryotic cells, they are also among the most complicated. As the following brief description indicates, the common idea that the amoeba is a simple primitive cell is incorrect. The amoeba most studied by cell biologists is *Amoeba proteus*. (*Proteus* comes from the Greek word *proteios* meaning first rank; the word *protein* has the same origin.) This cell has a length of one millimeter or more when stretched out into its streaming form, and has a volume 5000 times greater than the volume of a single yeast cell and 300,000 times that of *E. coli*.

The amoeba has no rigid cell wall, but its plasma membrane possesses a surface coat. Most if not all cells that lack a cell wall have a surface coat consisting of carbohydrate and protein molecules attached to the outside surface of the plasma membrane. Because of its thinness, the surface coat can be seen only with the electron microscope. The function of surface coats in eukaryotes is only partially understood. It probably contributes mechanical strength without counteracting the flexibility that the cell surface must have for amoeboid movement. In amoebae, it may also have a role in recognizing food organisms that are ingested by phagocytosis. In other kinds of eukaryotes, molecules of the surface coat serve as receptors for a wide variety of chemical signals from outside the cell.

The nucleus of the amoeba is extraordinarily large (an ellipsoid 30 μm in diameter) and contains dozens of nucleoli arranged in a layer immediately inside the nuclear envelope. A large number of nucleoli is not uncommon among protozoa, but in plant and animal cells there are usually only two to several nucleoli. The nucleus of the amoeba contains several hundred small chromosomes. The nuclear envelope is more elaborate than in yeast. On the inside of the envelope is a well developed honeycomb structure whose function is not known. This structure is called the *nuclear lamina*, a structure that is present in a less extensive and conspicuous form in other eukaryotic cell types. The lamina is a framework of proteins that stabilizes the nuclear envelope and serves for attachment of chromosomes. At the base of each honeycomb cavity of the amoeba nucleus is a pore in the nuclear envelope. As in the yeast cell, these pores are a route for the movement of materials between nucleus and cytoplasm. For example, subunits of ribosomal particles, which are produced in the nucleoli, move to the cytoplasm by passing through the pores in the envelope.

The cytoplasm of amoebae appears structurally more complex than that of the yeast cell. In addition to ribosomes and mitochondria a large number of membrane-bound vesicles collectively make up the *endoplasmic reticulum*, or ER. The endoplasmic reticulum is also present in the yeast cytoplasm but is far less conspicuous. The function of the endoplasmic reticulum will be discussed more extensively when animal cells are considered later in this chapter. The endoplasmic reticulum is involved in the synthesis of lipids and proteins, some of which are destined for transport to the cell surface, where they become part of the cell's surface coat.

The cytoplasm of amoebae contains several *Golgi complexes*. The Golgi complex, which is an aggregate of membranes, functions in processing, concentrating, and packaging into vesicles the macromolecules synthesized in the endoplasmic reticulum. These macromolecules will become part of the surface coat or be exported

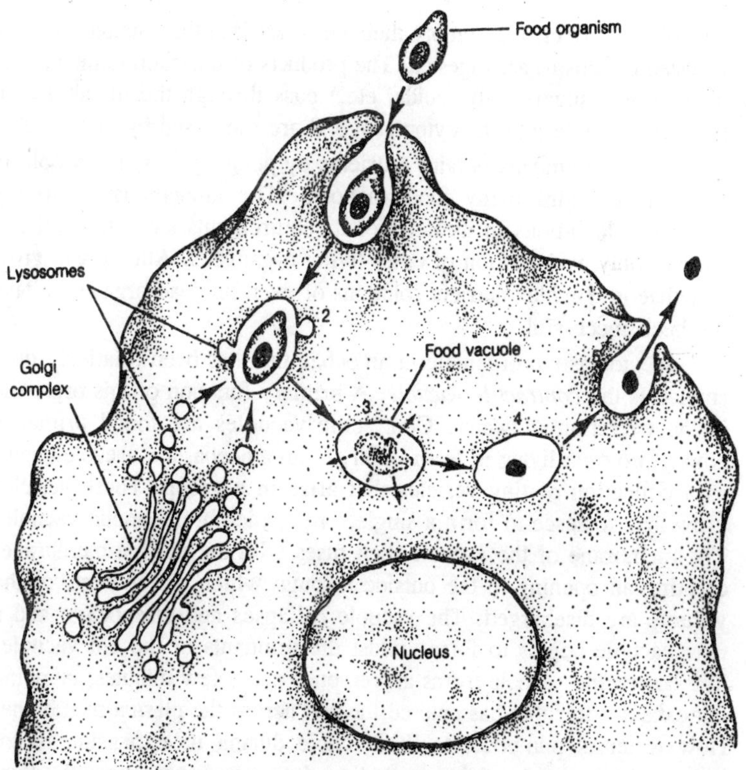

Figure 3.2 : Phagocytosis of a ciliated protozoan by an amoeba.

from the cell. In the case of amoebae, the surface coat must be continually renewed because of turnover of the surface that occurs during amoeboid movement and *phagocytosis*. Phagocytosis (derived from the Greek words *phagein* meaning to eat and *cytosis* meaning cell) is the process by which cells engulf particulate material. Amoebae acquire nutrients by ingesting organisms such as bacteria, algae, and other protozoa, which they then digest. In phagocytosis a portion of the plasma membrane surrounds the food organism to form a *food vacuole* within the cytoplasm.

Like the cytoplasm of most eukaryotic cells, the cytoplasm of amoeba contains numerous lysosomes. As in other cells, the lysosomal enzymes are released if an amoeba dies, and self-digestion, or autolysis, results. The lysosomes have another important function in amoebae and in other cells that ingest particulate matter. After ingestion of organisms by phagocytosis, lysosomes fuse with the newly formed food vacuole. In effect, they empty their enzymes into the vacuole, and the ingested organisms are digested. The products of digestion (amino acids, nucleosides, sugars, fatty acids, etc.) pass through the membrane of the food vacuole into the cytoplasm and are then used by the amoeba.

Since the amoeba obtains nutrients by phagocytosis, its cytoplasm naturally contains many food vacuoles. The amoeba is apparently extremely fastidious in its nutritional requirements since the cell can survive only by ingestion of other living organisms. Attempts to grow amoebae in complex nutrient solutions or even on dead organisms have not been successful.

A cytoplasmic organelle in amoebae that has been relatively little studied is the *contractile vacuole*. A primary function of this organelle is the excretion of water. Contractile vacuoles are found primarily among protozoa living in fresh water. In an amoeba, the vacuole begins as a collection of tiny vesicles that arise in a group. These vesicles grow and coalesce to form a single, large, membrane-bound vacuole. The membrane of the vacuole then fuses with the plasma membrane, creating an opening to the outside through which the contents of the vacuole are discharged. The vacuole collapses and disappears, and a new vacuole begins to form amidst the remnants of the old vacuole. The contractile vacuole forms and empties every few minutes, pumping out excess water that has entered the cell across the plasma membrane. How water is pumped into the vacuole during its expansion is not known, but mitochondria are believed to participate because they aggregate on the surface of the vacuole, perhaps providing the necessary chemical energy for pumping water into the vacuole.

Like other eukaryotes, the cytoplasm of amoebae contains fibrous elements known as microtubules and microfilaments. These are not easily observed in amoeba and will be discussed later in this chapter in animal cells, where they are more conspicuous.

The amoeba has been useful in studies of the interaction between the nucleus and the cytoplasm because the cell is large enough for microsurgical operations. The amoeba was one of the first cell types to be cut into two parts with a fine needle, one part containing the nucleus and the other containing only cytoplasm. The nucleated half-cell regains its lost cytoplasm by growth and subsequently resumes proliferation. The enucleated half-cell lives for two to three weeks, during which time its metabolic activities decline steadily. The enucleated half-cell is unable to maintain its structural integrity and eventually disintegrates. By comparing the activities of nucleated and enucleated half-cells, cell biologists first learned about the roles of the nucleus and cytoplasm in various cell functions.

The amoeba is one of the few types of cells in which a nucleus can be transplanted from one cell to another. Essentially, the operation consists of inserting an extremely fine glass needle into an amoeba and pushing the nucleus through the plasma membrane into an adjacent cell.

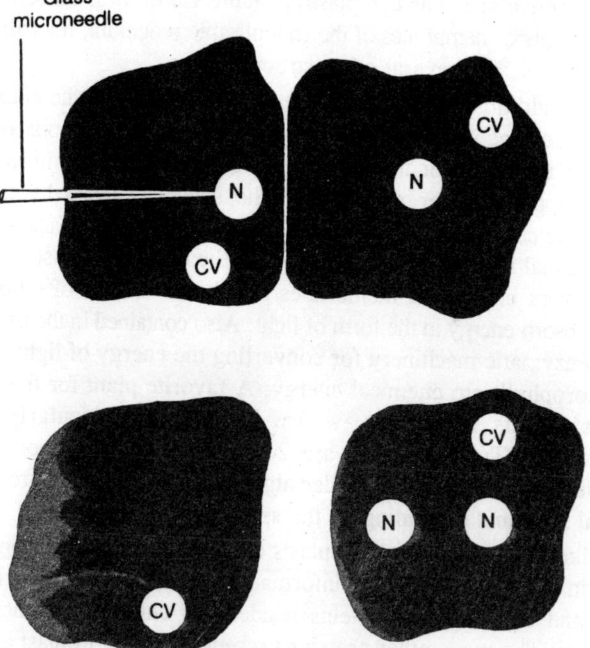

Figure 3.3 : Nuclear transplantation in amoebae.

The needle must be moved by an instrument called a *micromanipulator* because even the surest hand is by itself too unsteady. The micromanipulator transmits movements of the hand to a tool like a microneedle but reduces the magnitude of the hand movement. For example, movement of one centimeter of the hand in a given direction is reduced to movement of a few micrometers (10^{-4}cm) of the microneedle. Nuclear transplantation experiments in amoebae provided the first evidence that certain kinds of proteins constantly migrate back and forth between the nucleus and cytoplasm.

Plant Cells

Most plant cells are enveloped by a thick, rigid wall that gives the cell its shape and protects it both from mechanical injury and from osmotic swelling. The wall is made up of cellulose fibers embedded in a matrix of other polysaccharides and proteins. Materials that make up the cell wall are synthesized inside the cell, then packaged just inside the plasma membrane, released from the cell, and deposited in the cell wall.

Plant cells, like all eukaryotic cells, contain a nucleus enclosed by a nuclear envelope. The cytoplasm contains ribosomes, mitochondria, a Golgi complex, membranes of the endoplasmic reticulum, microfilaments, microtubules, and several other structures.

Two additional structures present in plant cells are the *chloroplasts* and the *central vacuole*. The chloroplast is a membranous organelle that, like the mitochondrion, has a surrounding outer membrane and a closely apposed inner membrane. The space enclosed by the inner membrane contains a complex solution of enzymes and stacks or layers of additional membranes that largely fill the organelle. Associated with these layers of internal membranes are the molecules of chlorophyll, which absorb energy in the form of light. Also contained in the membranes is the enzymatic machinery for converting the energy of light absorbed by chlorophyll into chemical energy. A favorite plant for the study of photosynthesis is spinach, whose leaves are particularly rich in chloroplasts. Typically there are 20 to 30 chloroplasts per cell in a plant leaf, although in unicellular algae the number varies from one to several thousand depending on the species examined.

Like mitochondria, chloroplasts contain DNA and ribosomes. The DNA in chloroplasts contains information for the synthesis of ribosomal RNA and some of the proteins needed for chloroplast function and structure. The many other proteins present in the chloroplast are coded for by DNA in the cell nucleus. Through interactions with the cytoplasm and nucleus that are not yet well understood, the function, growth, and reproduction of chloroplasts are closely coordinated with the activities and growth of the cell.

The central vacuole of the cytoplasm in a plant cell contains water, sugars, organic acids, proteins, inorganic ions, and pigments. Many of these are waste and storage products that accumulate during the life of the cell.

The nuclei of plant cells closely resemble the nuclei of other kinds of eukaryotes. Nuclei contain the chromosomes and one or more nucleoli and are enclosed by an envelope penetrated by numerous pores. As in other cells, the nucleus produces all the RNA present in the cytoplasm, except for the small amount of RNA produced in mitochondria and chloroplasts. These cytoplasmic organelles account for less than one percent of the total RNA synthesis of the cell.

Although individual cells from a multicellular plant can now be cultured and grown in a nutrient medium of known constitution, they have been used less often than animal cells in research (except in the study of photosynthesis). Research on plant cells has recently become intense, particularly because of the potential for genetic engineering to introduce major new properties into plants of enormous economic importance. For molecular analyses of photosynthesis and other activities, plant cells are usually broken open by mechanical means, such as grinding or agitation in a kitchen blender; the chloroplasts or other organelles are then separated from other cell constituents by sedimentation in a centrifuge, described later in this chapter.

Plant and algal cells (and the photosynthetic bacteria) are uniquely important, because almost all the energy needed to sustain life on Earth is derived from the photosynthetic capture of radiant energy from the sun and because these photosynthetic organisms produce much of the oxygen in the atmosphere.

Animal Cells

Complex animals such as mammals are composed of about two hundred different types of cells. Each type has become specialized during embryological development to perform a particular function in one of the many tissues of the animal. These specialized differentiated cells include blood cells, liver cells, kidney cells, skin cells, bone cells, germ cells, and many others. Some of these cells can grow and divide in culture vessels if they are provided with a complex nutrient medium containing amino acids, inorganic ions, sugars, vitamins, and many other substances.

As in amoebae, the surface of the plasma membrane of cultured animal cells is coated with a thin layer of *glycoproteins*—proteins with various kinds and numbers of sugar molecules attached to them. One function of the surface is to provide the means by which cells can adhere to one another in tissues. As in other cells, the plasma membrane consists of a lipid bilayer with atached proteins. Many of the membrane-associated proteins function in the absorption of nuttrients from the medium, and others serve as receptors for signals such as hormones.

The cytoplasm of animal cells contains ribosomes, endoplasmic reticulum, Golgi complexes, mitochondria, lysosomes, lipid droplets, and five structures generally present in eukaryotic cells but not yet described: microtubules, microfilaments, intermediate filaments, microbodies, and centrioles.

The two fiber structures, *microtubules* and *microfilaments*, that generally occur in eukaryotes are particularly prominent in some kinds of protozoa and in animal cells. Microtubules are long, hollow structures with a diameter of 30 nm and a wall thickness of about 8 nm. Microtubules are assembled from subunit protein molecules called *tubulin*. Microtubules occur in bundles or singly and impart stiffness and form to a cell. They thus provide a structural framework or *cytoskeleton*. For example, long extensions of cytoplasm, such as the *axons* of nerve cells, contain prominent longitudinal bundles of microtubules. Microtubules are also the main structural element in cilia and flagella. Except for those in cilia and flagella, microtubules are labile structures that disassemble into tubulin subunits and reassemble into new microtubules during changes in cell shape. A particularly noteworthy example of this process of assembly and disassembly occurs during cell division. The *mitotic apparatus*, which serves to distribute the chromosomes to the daughter cells, consists in large part of microtubules that assemble from tubulin sub-units and disassemble once mitosis is complete.

Another fibrous element found in all eukaryotic cells is the microfilament. These are long, thin fibers, 9.5 nm in diameter, made of proteins. Microfilaments are often conspicuously present in bundles in animal cells in culture. Individual microfilaments are sometimes anchored to the plasma membrane, either at their ends or along their lengths. They are made largely of *actin*, one of the major proteins in the contractile apparatus of muscle cells. Actin is one of the most abundant proteins in eukaryotic cells. In nonmuscle cells actin microfilaments are important in amoeboid movement and other kinds of cell movement.

A third type of fiber structure is the *intermediate filament*. These are present in most if not all eukaryotes but are particularly numerous in certain kinds of animal cells. A variety of kinds of intermediate filaments has been found, each composed of a different kind of protein subunit, and often several kinds of proteins in a single filament. For example, skin cells contain intermediate filaments composed of proteins called keratins; their presence in large numbers gives skin cells their special mechanical toughness. Nerve cells are rich in intermediate filaments called neurofilainents composed of several other kinds of proteins. Muscle cells contain their own kind of intermediate filaments. In every case, however, the role of intermediate filaments seems to be a passive, structural one; they form mechanical connections between different parts of the cytoplasm. Microtubules, microfilaments, and intermediate filaments are the major elements making up a fibrous network in the cytoplasm of eukaryotic cells known as the cytoskeleton. The cytoskeleton, which has the major role in giving a cell its particular shape. The roles of microtubules and microfilaments in cell motility are discussed.

Centrioles are present in animal cells and in many kinds of unicellular eukaryotes. A centriole is a cylindrical structure about 0.15 μm in diameter and 0.3 to 0.5 μm long. It is made up of nine triplets of microtubules in a pinwheel-like arrangement that runs the length of the centriole. A matrix of unidentified, amorphous material surrounds the triplets of microtubules. Centrioles almost always occur in pairs. The two centrioles in a pair lie with their long axes at right angles to each other. Centrioles have traditionally been believed to play a role in mitosis although their importance is uncertain, since many cells have no centrioles. For example, plant cells, amoebae, and many algae lack centrioles but nevertheless divide mitotically. In cells that are preparing to divide, two pairs of centrioles are present, and during cell division each daughter cell receives one pair.

Microbodies are small vesicles (usually 0.5 to 2 μm in diameter) composed of a granular matrix enclosed by a single membrane. Many different enzymes are present in microbodies. Most of these enzymes are involved in the oxidation of organic molecules, for example, amino acids and fats by molecular oxygen. The makeup of enzymes in microbodies varies somewhat from one type of cell to another, which has at times raised doubts about classifying them as a single entity. However, one common feature of microbodies is the production of hydrogen peroxide (H_2O_2). Because of the formation of hydrogen peroxide as a result of oxidative processes in microbodies, they came to be known as *peroxisomes*. Hydrogen peroxide is a strong oxidizing agent and potentially harmful to the cell, but it is rapidly destroyed by an enzyme *catalase* within the peroxisome to yield water and oxygen:

$$2H_2O_2 \rightarrow 2H_2O + O_2$$

In cells of plant seedlings rich in stored fat, for example peanut and cucumber seedlings, vesicles similar to microbodies and known as *glyoxysomes* convert fats to simpler molecules that can then be used in other parts of cell metabolism, including the generation of usable chemical energy.

The nucleus of an animal cell, like that in other eukaryotes, has a nuclear envelope with numerous pores, one or more nucleoli, and a network of chromosomal fibers. During the period between cell divisions, it is impossible to discern complete chromosomes in the nucleus because each chromosome is an extremely long, twisted thread of DNA and protein jumbled into a network with threads of other chromosomes. However, during mitosis the sinuous thread that comprises each chromosome condenses into a short, rod-shaped object, the mitotic chromosome.

Cells within a tissue of an animal usually retain a definite shape that results in part from structural properties acquired during differentiation and in part from physical constraints imposed by adjacent cells. Because such influences are greatly diminished when cells are removed from a tissue and grown in culture vessels, cultured cells are constantly moving and changing shape, though usually at rates too low to be detected by eye in a microscope. Movement is normally detected by time-lapse photography.

The electron microscope reveals that the overall form of an animal cell is considerably more complex than the light microscope would indicate. The view is particularly dramatic in a scanning electron microscope, which provides a high-resolution image of the surface of an object rather than showing its internal structure. A scanning electron microscope photograph of a cell in culture reveals that the cell surface is covered with a variety of hairlike projections and blebs of different sizes. Although the significance of these structures is not well understood, the projections give the cell a much greater total surface area than had previously been assumed. The periphery of the cell also shows variable form. Ruffles at the edge of the cell identify regions of active movement of the cytoplasm. Along with this movement, tiny invaginations of the cell surface form vacuoles that detach from the inner surface of the cell and drift into the cell interior. This process of surface invagination and vacuole formation, which occurs in eukaryotes that lack a rigid cell wall, is called *pinocytosis* or *cell drinking*. Cells can take in large amounts of nutrients by pinocytosis. Some viruses also gain entry into cells by inclusion in pinocytic vacuoles. Overall pinocytosis is similar to phagocytosis. Pinocytosis is the ingestion of fluid and phagocytosis is the ingestion of particulates such as bacteria. Pinocytic vesicles are generally much smaller than phagocytic vesicles.

SPECIALIZATION OF CELLS FOR PARTICULAR FUNCTIONS

The cells that make up the many tissues of an animal are, for the most part, differentiated to perform various specialized functions. To carry out a specialized function a cell may develop specialized structures in the cytoplasm: for example, contractile filaments develop in muscle cells; axons and dendrites in nerve cells; flagella in sperm cells; and cilia on some kinds of epithelial cells. Other cells acquire the ability to synthesize specialized chemical products such as digestive enzymes, which are produced in cells of salivary glands and the pancreas. Glandular cells produce hormones. Cells in bone, cartilage, and connective tissue synthesize and release collagen, a protein that gives these tissues their mechanical strength.

The specialized biochemical activities of particular cells are frequently reflected in the structural organization of the cytoplasm and nucleus. A liver cell, or *hepatocyte*, has a high concentration of ribosomes in the cytoplasm and a nucleolus that is especially active in ribosome production. These structures are present because liver cells synthesize large amounts of protein (e.g., the blood protein serum albumin) that is released from the cell to become a major constituent of the blood. Most of the ribosomes in a liver cell are bound to membranes of the endoplasmic reticulum (ER), which is the characteristic location of ribosomes working to synthesize proteins destined for export from the cell via the Golgi complex. In contrast, proteins retained by the cell in the cytosol are synthesized mostly on ribosomes that are not attached to the ER. That portion of the ER to which ribosomes are attached is called the rough ER because of its studded appearance in electron micrographs. The cytoplasm of pancreas cells is dominated by the rough ER, because these cells synthesize digestive enzymes (proteins) that are released from the cell and then carried by the pancreatic duct to be emptied into the intestine.

In all cell types, a part of the ER lacks attached ribosomes and is called the smooth ER. The membranes of the smooth ER contain, among other things, enzymes that catalyze the synthesis of steroids. Therefore, the smooth ER is prominent in cells that synthesize and secrete steroid hormones, for example, some cells in the testis and those in the cortex of the adrenal gland. In skeletal muscle cells the molecular composition of the smooth ER is modified to play a major role in controlling muscle contractions. Membranes of both the smooth and rough ER synthesize lipids; they also join lipids to proteins to form lipoproteins, some of which are destined to become part of the plasma membrane. Built into the membranes of the smooth and rough ER of hepatocytes are enzymes that detoxify drugs such as barbiturates,

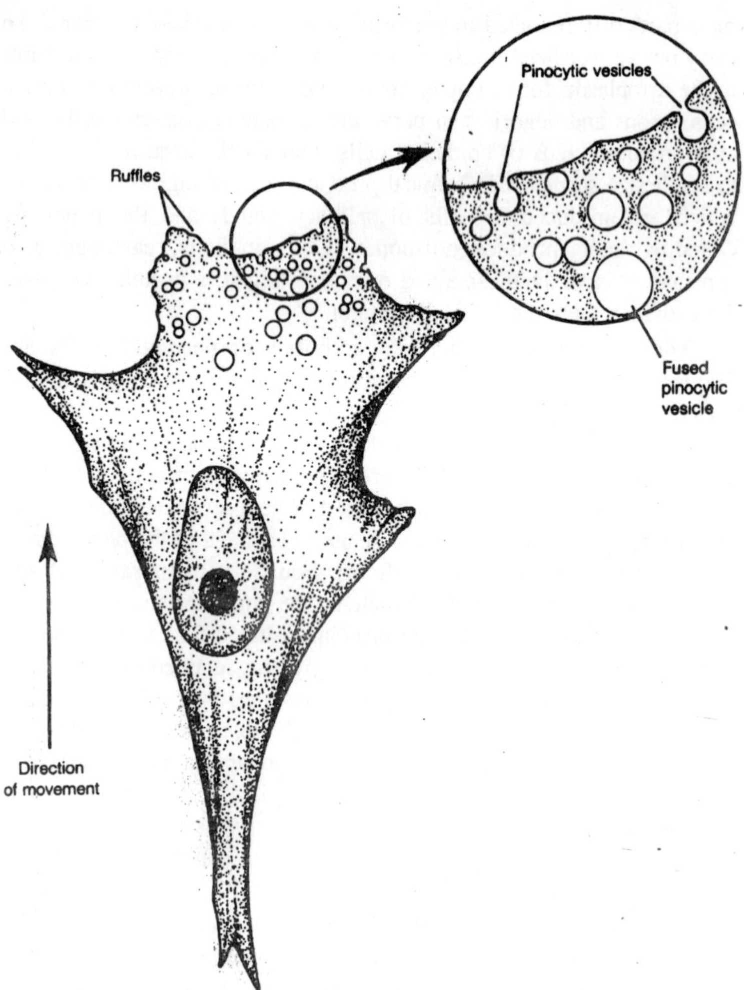

Figure 3.4: Diagram of pinocytosis. The advancing edge of an animal cell is marked with sheetlike projections called lamellipodia *or ruffles.*

tranquilizers, and ethyl alcohol, as well as environmental pollutants such as insecticides, herbicides, dyes, and some food preservatives.

A SURVEY OF CELL FUNCTIONS

Essentially all cell activities are devoted to one or more overall functions; cell self-maintenance, cell motility, cell growth and reproduction, and the specialized work of differentiated cells. In effect, these functions encompass everything cells do.

Self-maintenance consists of the continuous synthesis of new molecules to replace those that become damaged. Within a cell, enzymes, the components of ribosomes, membranes, and other parts are constantly subject to damage from the highly energetic motions of ions and molecules. The tendency of cytoplasmic components to become damaged is particularly evident when the nucleus of a cell is removed. Without a nucleus, the cell is deprived of the genetic information required for protein synthesis; hence, replacement of its cytoplasmic parts cannot continue. In the first hours after enucleation, the cell continues to synthesize proteins and lipids, mobilizes energy, takes in nutrients, retains motility, and constantly changes shape. By 12 hours after enucleation, the rate of protein synthesis in the cell has declined to a fraction of its original rate; by 18 hours later, protein synthesis is no longer detectable. Experimentally enucleated animal cells live as long as several days, steadily decreasing in size, until self-maintenance fails completely and the cell disintegrates.

Self-maintenance also includes exchanges between the cell and its environment, particularly the taking in of nutrients that supply the cell with sources of chemical energy and raw materials to use in synthesizing replacement macromolecules. Self-maintenance is a housekeeping operation consisting of activities necessary for the functional and structural health of a cell. The cost of such housekeeping is met by the continual intake of sources of chemical energy and of nutrients along with the concomitant excretion of metabolic end products.

Cell motility takes several forms: cytoplasmic streaming, movement of organelles and particles inside cells, active changes in cell shape, amoeboid movement, propulsion by cilia and flagella, and muscle contraction. All forms of motility appear to be based on a particular set of proteins whose interactions with one another inside the cell produce movement. All forms of movement require chemical energy.

Cell growth and *reproduction* consist of those activities by which a cell increases the amount of its metabolic machinery and structural components, which are eventually distributed to two daughter cells at division. One of the most remarkable aspects of cell growth is the integration of the synthesis of thousands of different proteins and other molecules. A large and complex set of interactions, about which we still know very little, coordinates all these activities. Thus in a growing eukaryotic cell, chromosomes, ribosomes, mitochondria, endoplasmic reticulum, plasma membrane, and other structures, as well as thousands of proteins and other macromolecules, all increase in a precisely controlled and coordinated fashion. Growth prepares the cell for reproduction, which is the separation of the cell into two daughters by cell division. Cell division consists of two closely coordinated events: one is nuclear division, called *mitosis* or *karyokinesis*, which accomplishes the precise distribution of chromosomes to the daughter cells; the other is *cytokinesis*, the dividing of the cytoplasm to form two cells.

Cell self-maintenance is necessary for the survival of every species. It is perhaps less immediately evident that cell growth and reproduction are also stringent requirements for species survival. Without cell proliferation, no species of organism, either unicellular or multicellular, could reproduce to replace those individuals of its kind that die.

Unicellular organisms have special elaborate structures to carry out specialized activities, for example cilia, which are organelles of motility. However, this kind of differentiation is limited and does not approach the diversity, flexibility, and scope of the differentiation that occurs in multicellular plants and animals. The evolutionary success of multicellular organisms resides in their ability to produce differentiated cells to carry out specialized functions. The nature of differentiation is apparent in the development of an organism such as a human. Development begins with the formation of a single cell, the fertilized egg cell or zygote. Undergoing a rapid series of cell divisions, the human zygote consists of 64 cells by the end of 3 days. Most of these are destined to form the placenta, and a few to form the embryo. Only 21 days after fertilization the embryo consists of hundreds of thousands of cells that have differentiated into dozens of types. By this early point in development the heart has formed and begun to contract rhythmically, pumping newly formed blood cells through a rapidly developing system of arteries and veins. At birth the fetus is comprised of 10^{13} cells consisting of about two hundred or more distinctly dif-ferentiated types.

Cell growth and reproduction, as well as differenti-ation, continue to take place in the adult organism to replace cells that wear out. For example, each adult human has about 2.5×10^{13} red blood cells. The aver-age red blood cell lives about 120 days, or 10^7 seconds. Therefore, the body must produce 2.5×10^{13} new red blood cells every 10^7 seconds, or 2.5×10^6 new red blood cells every second. In order to maintain this rate of production among those cells in the bone marrow that give rise to red blood cells, 2.5×10^6 cells must divide each second. Although few tissues in an organ-ism produce cells at this rate, the total number of cells produced and differentiated in all of the many tissues of the human body probably exceeds 10^7 per second.

A SURVEY OF CELL NUTRITION

All of the activities carried out by a cell are fueled by an environmental supply of energy and depend on an environmental source of carbon, hydrogen, oxygen, nitrogen, phosphate, sulfate, and other components from which the molecules of the cell can be built. How these requirements are fulfilled varies greatly from one kind of cell to another. However, despite the variability, cells can be adequately characterized with only a few groups depending, first, on the chemical form of carbon they require, and second, on the form of the energy they obtain from the environment.

Those cells that can build all their organic molecules from only carbon dioxide (CO_2) are called *autotrophs* (derived from the Greek *autos* meaning self and *trophe,* nurturing). Photosynthetic cells and some bac-teria make up the autotrophs. Cells that must fulfill their carbon requirement by taking in chemically re-duced carbon in the form of organic molecules (for example, sugars) are called *heterotrophs* (Greek *hetero,* other). Most nonphotosynthetic bacteria, fungi, protozoa, and animal cells are heterotrophs. The degree of heterotrophy varies over a wide range; some cells, such as *E. coli,* can utilize a single organic compound, glucose, for building all of the complex molecules within the cell. Other heterotrophs require a broad spectrum of organic molecules; for growth the human cell requires ten preformed amino acids, sugar, numerous vitamins, and several fatty acids.

Cells are classified as *phototrophs* and *chemotrophs* depending upon how they obtain energy from the environment. Phototrophs are those cells that use the energy of light in order to synthesize sugar by photosynthesis. Chemotrophs are those that must derive energy through chemical reactions in which substances, usually carbohydrates, are broken down into simpler substances.

Almost all phototrophs are autotrophs; that is, they obtain energy by photosynthesis and use CO_2 as a source of carbon from which to build all their organic molecules. A few phototrophs require one or another organic molecule, and hence are called *photoheterotrophs*. For example, the alga *Euglena* requires the vitamins thiamine and $_{B12}$. Almost all chemotrophs are heterotrophs, requiring organic molecules as building blocks and digesting organic molecules to obtain chemical energy. Ultimately, chemotrophs-heterotrophs, such as animals, must ingest phototrophs-autotrophs (plants) to obtain a supply of organic molecules from which to obtain both building blocks and energy.

This brief review of cell functions and structures will be followed, in later chapters, by an expanded view of cell properties. Explanations of how cells are built and how they work, however, presuppose some knowledge of cell chemistry. The discussion of cells therefore continues in the next chapter with a review of the chemical components of and chemical interactions that take place in cells.

A BRIEF COMMENT ON METHODS IN CELL RESEARCH

This chapter shows that many methods have been devised for the study of cell function and structure. The oldest tool in cell research is the microscope. Microscopy has had a long history of development and improvement and continues to be invaluable in the study of cell structure. *Microsurgery* has also been mentioned as a means of analyzing cell functions, for example, the transplantation of a nucleus from one cell to another, as in amoebae. Other research tools and techniques are described, as appropriate, in succeeding chapters.

Modern cell research relies heavily on hundreds of techniques of biochemical analysis to determine the molecular nature of cell functions and structure. Biochemical analysis often requires large amounts of a purified cell component and therefore depends on the ability of the cell biologist to lyse cells and fractionate them into their structural parts. Cell fractionation continues to be central to much of cell research and is briefly described below. Indeed, identification of the biochemical composition and function of many cell parts and organelles was made possible by cell fractionation.

Cell Fractionation

Cell fractionation is the breaking open of cells and separation of the parts into pure fractions. The technique generally requires a large number of cells. Convenient sources are cells of plant and animal tissues and animal cells grown in culture, although bacteria, protozoa, algae, and fungi are frequently used. The breaking open of cells is called *lysis* or *homogenization* and is usually accomplished by mechanical disruption of the plasma membrane with an homogenizer. Sometimes the plasma membrane is dissolved with a detergent solution. A homogenizer consists of a tube and a close fitting pestle. The cells are placed in the tube in an appropriate solution of inorganic ions and low-molecular-weight organic molecules (e.g., sucrose) that will maintain as far as possible the functional and structural properties of the cell parts once the cells are broken open. The pestle is inserted into the tube and rotated as it is drawn in and out of the tube. The motion of the pestle creates a shearing action that breaks open the cells. Some cell structures such as mitochondria, chloroplasts, lysosomes, microbodies, nuclei, and ribosomes remain largely intact during homogenization while the endoplasmic reticulum, Golgi complex, and plasma membrane are usually fragmented. (However, special methods have been devised to isolate the Golgi complex and plasma membrane intact.) The endoplasmic reticulum is broken into small fragments that spontaneously form small sealed vesicles called *microsomes*. The major structural components can then be separated into pure fractions by centrifugation. Separation is possible because the velocity with which a particular cell structure sediments in a centrifugal field depends on its size, density, and shape.

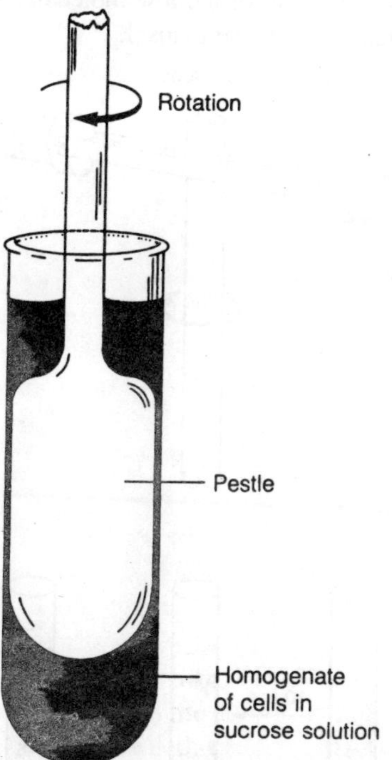

Figure 3.5 : Different structural components present in a cell homogenate can be separated by centrifugation.

Although there are various ways to fractionate a cell homogenate (lysate), the following method, known as *differential centrifugation*, illustrates the principle. A cell homogenate is placed in a centrifuge tube, which is then centrifuged. Nuclei are usually the largest intact structures in the homogenate (and are denser than water) and hence will sediment most rapidly and will be the first to accumulate at the bottom of the tube, forming a pellet of nuclei. At this point, the centrifugation is stopped and the cell homogenate (the supernatant) removed. The nuclear pellet is left behind, while the cell homogenate is placed in a second tube and centrifuged again, this time at higher centrifugal speed. This second centrifugation causes the next largest intact structures (usually the mitochondria) to sediment into a pellet. The process is repeated several times, with increasingly stronger centrifugal forces to obtain successively smaller structures. The final structures to sediment into a pellet are the ribosomes. The supernatant that remains after sedimentation of the ribosomes is called the *soluble fraction*. It contains many enzymes as well as all of the low-molecular-weight components (sugar, amino acids, etc.) of the cytosol.

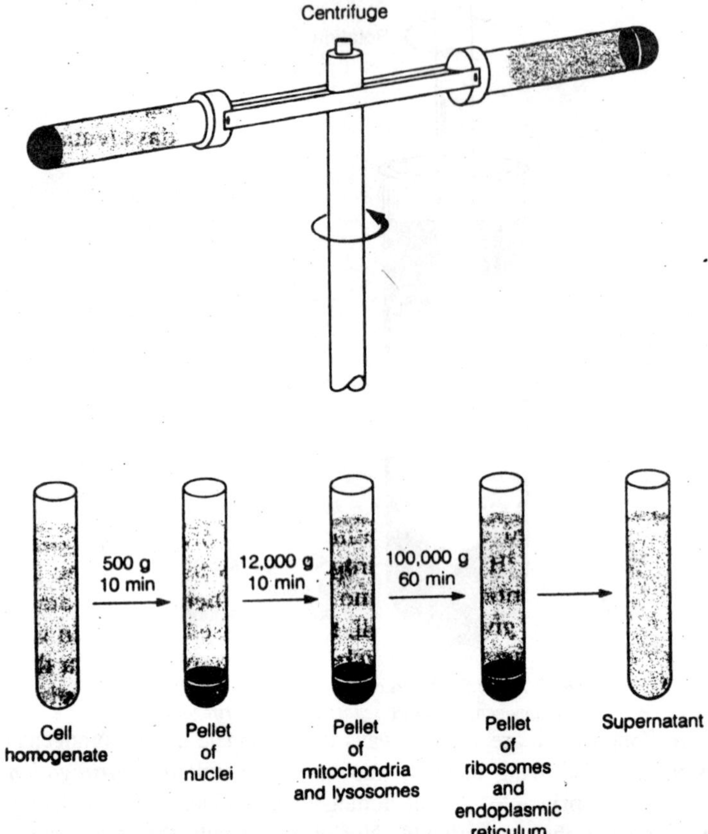

Figure 3.6 : Different structural components present in a cell homogenate can be separated by centrifugation.

Once organelles have been obtained as pure fractions they can be analyzed for biochemical makeup and function. Some organelles can be subfractionated; purified mitochondria can be broken and separated into the inner and outer membrane parts, and nucleoli can be isolated from purified nuclei. These separations enable further refinement to be made in the analysis of function and structure.

Radioactive Isotopes in Cell Biology

Biochemical analysis of the functions and structures of cells is aided enormously by the use of radioactive isotopes as tracers of cell chemistry. They permit precise and detailed measurements that would other-wise be impossible. An isotope is a form of an element in which an extra neutron is present or a neutron is missing from the atomic nucleus. An abnormal number of neutrons often makes the atomic nucleus unstable, and they disintegrate, emitting radiation. For example, hydrogen contains one proton and no neutrons in its nucleus. Deuterium is an isotope of hydrogen containing one proton and one neutron. Tritium has one proton and two neutrons. Deuterium is stable but tritium is unstable; one of the neutrons splits into a proton and an electron. The proton remains in the atomic nucleus but the electron, called a β particle, is ejected from the tritium atom and can be detected with a radioactivity counter or with photographic film.

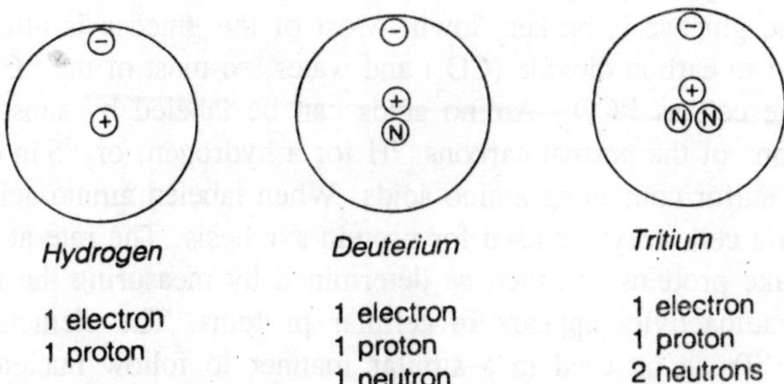

Figure 3.7: Hydrogen exists in three isotopes, differing in the number of neutrons in their atomic nuclei.

Four radioactive isotopes widely used to investigate cell chemistry are tritium (^{3}H), radioactive carbon (^{14}C), radioactive phosphorus (^{32}P), and radioactive sulfur (^{35}S). Properties of these isotopes and kinds of molecules into which they become incorporated are listed in Table 1-2. Illustrations of how they are used to study cell functions and structures will appear frequently in the chapters that follow.

Each kind of radioactive isotopes decay at a characteristic rate called its *half-life*, the time required for half of the radioactive atoms present to disintegrate. The half-life of ^{32}P is 14.3 days, which means that starting with a given amount of the isotope, 50 percent of the atoms will have disintegrated (decayed) in 14.3 days; half of the remaining will decay in the next 14.3 days, and so on. By 42.9 days (equal to 3 half-lives) 87.5 percent of the atoms will have decayed.

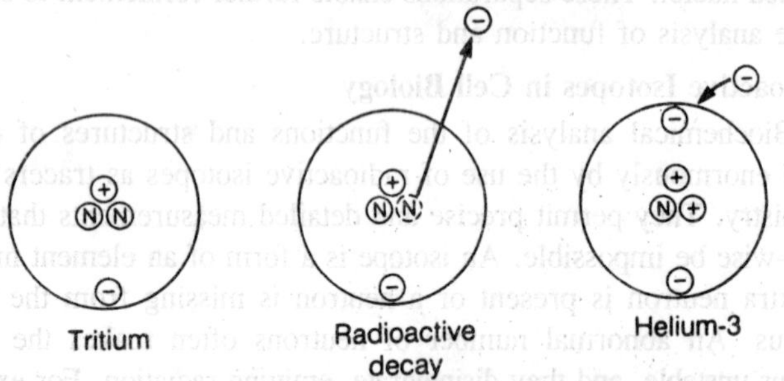

Figure 3.8: Dercay of tritium. Tritium is an unstaboe isotope of hydrogen. During radioactive decay it emits an electron.

Because they can be substituted for normal atoms in organic molecules, ^{3}H, ^{14}C, ^{32}P, and ^{35S} are particularly useful in cell research. For example, it is possible to produce glucose molecules in which one or more carbon atoms is replaced by ^{14}C. The radioactively labeled glucose can be given to cells and the metabolic fate of glucose determined by observing which molecules in the cell become radioactive when the glucose is broken down. Most of the glucose is ultimately degraded to carbon dioxide (CO_2) and water, so most of the ^{14}C atoms leave the cell as $^{14}CO_2$. Amino acids can be labeled by substituting ^{14}C for one of the normal carbons, ^{3}H for a hydrogen, or ^{35}S in one of the two sulfur-containing amino acids. When labeled amino acids are given to a cell, they are used for protein synthesis. The rate at which cells make proteins can then be determined by measuring the rate at which radioactivity appears in cellular proteins. The elements ^{3}H, ^{14}C, or ^{32}P can be used in a similar manner to follow nucleic acid synthesis and breakdown.

The amount of radioactivity in a cell in the form of labeled molecules can be measured electronically or photographically. Electronic measurements are typically done with a scintillation counter. In scintillation counting, radioactive cells, cell fractions, orcell extracts are mixed into a solution containing a molecule that will scintillate, that is, emit light when it absorbs a β particle. Each decay produces a pulse of light that can be counted. Furthermore, the intensity of the light pulse is proportional to the energy of the decay and can be used to identify the isotope that is decaying. This feature allows one to use two different isotopes simultaneously to label different substances (e.g., proteir and DNA).

Alternatively, radioactivity in cells or macromole cules can be detected by a photographic film. Radioactivity causes exposure of the film in the same way tha light does. This technique, called *autoradiography* has the advantage of showing precisely where the ra dioactivity is located in the cell.

4

Aquatic Environment of Cell

A cell must always possess direct or indirect water frontage, inasmuch as all cell reactions occur in aqueous solutions. In simpler forms of life th is ·is acfiieved by direct contact with environmental water. The more complex plant or animal may become covered with an outer coating impervious to water, but each of its cells has contact with the aqueous environment of the internal medium.

Water makes up 80 per cent of the cell. When the cell loses its water, life is suspended or extinguished. Activity reappears in desiccated, dormant cells when water again becomes available. The cells of some dried mosses, lichens, rotifers, tardigrades and nematodes resist unfavourable conditions for years, only to resume activity quickly when moistened, provided other conditions are favourable. Many protozoa encyst and some bacteria iporulate when deprived of water. Such dried resistant stages are especially characteristic of fresh water and terrestrial organisms which are subjected to seasonal drying.

Many properties of water make it suitable not only as the main constituent of the living cell, but also as the main factor in the cell environment.

Water as a Temperature Stabiliser

Among the most important properties of water are its *temperature-stabilising* properties, especially its high specific heat, its high heat of vaporisation and its high heat of fusion.

The *specific heat* of a 'substance is defined as the number of calories required to raise the temperature of 1 gram of that substance 1°C. The specific heat of water is 1.0, a value exceeded only by that of liquid ammonia.

The *latent heat of vaporisation* is the number of calories required to change 1 gram of liquid to vapor. For water at 100° C. this is 539 calories, one of the highest values known. Close to the heat of vaporisation of water is that of hydrofluoric acid (360 calories); that of ammonia is less.

The *heat of fusion* is the number of calories required to convert 1 gram of solid at the freezing point to 1 gram of liquid at the same temperature. For water it is 80 calories and for ammonia, 108 calories per gram. Although the heat of fusion of water is less than that of ammonia, it is greater than that of most other organic or inorganic materials tested.

How do these properties of water provide a better control of temperature on the surface of the earth as well as in the external and internal environment of the cell? This can be illustrated best by a number of examples. Radiant energy of the order of 10^{15} cal./km^2 a year falls on the surface of the earth at the equator. This is equivalent to the energy liberated by the fission of 100 kg of uranium-235 or by the burning of 700,000 tons of coal.

If a dry land mass made up of rock and soil with a low specific heat were to absorb this much radiant energy, it would rise to a lethal temperature by day and fall below the freezing point of water by night, as it apparently does on the moon. However, the water on earth and water vapor in the atmosphere of the earth absorb some of the incoming radiation in sunlight, preventing overheating in daylight. Some of the sunlight which gets through the atmospheric water vapor "blanket" is absorbed and heats the land masses. As the heat is reradiated from the land by night, the atmospheric water vapor absorbs the outgoing infrared rays, preventing the earth from cooling at night as rapidly as it otherwise would. This so-called "greenhouse effect" of the atmospheric water vapor, the moderation of the temperature of the land masses, is of special importance to life.

Because of the high specific heat of water, the temperature of the great bodies of water covering the earth's crust is increased much less than that of the land. A body of water can absorb or lose a considerable amount of heat without much change in its temperature; that is, its heat capacity is high. The *heat capacity* of a body may be defined as the number of calories required to raise the temperature of the entire body

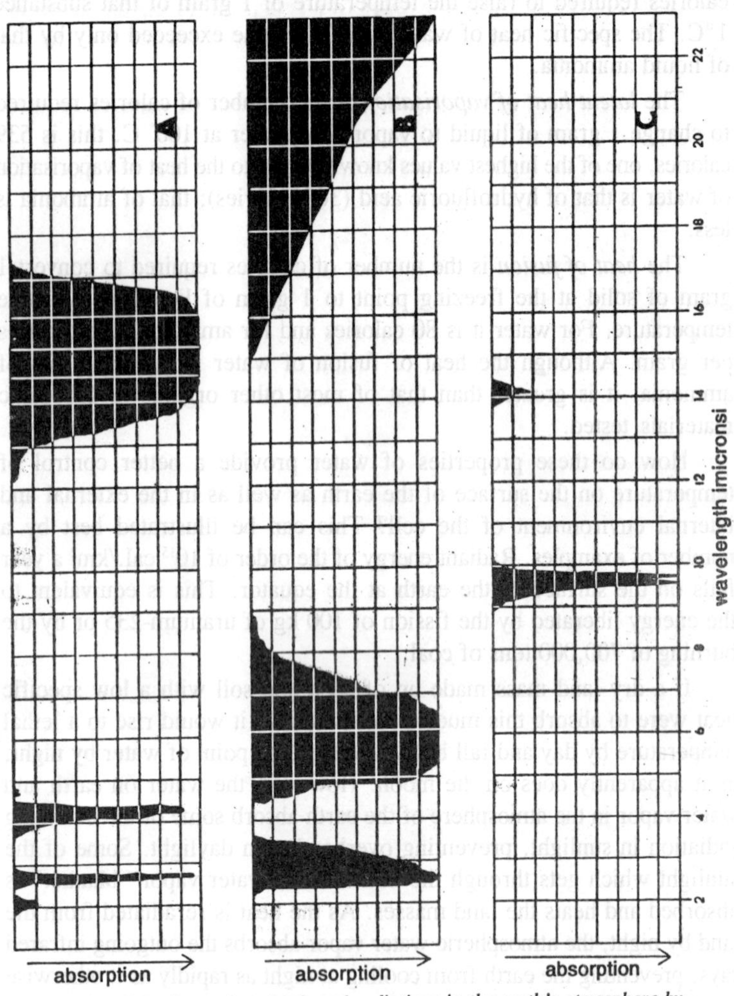

Figure 4.1: Absorption of infrared radiations in the earth's atmosphere by carbon dioxide (A), water vapor (B) and ozone (C).

1 °C. The larger a body is and the higher its specific heat, the greater is its heat capacity. The vastness of the oceans makes them of especial importance as thermostats resisting temperature change. Consequently, the climate of land masses is tempered by nearby water.

The oceans moderate the temperature on earth in yet another way. Because of the high heat of vaporisation of water, a tremendous amount of the sun's heat is absorbed during evaporation from the vast surface of the oceans. The vapor holds the heat until the warm, water-laden air is chilled in a cooler part of the earth. Then the vapor condenses and liberates the heat, thereby counteracting the coolness of the region in *which* precipitation occurs. The water-laden air may move far from its original source before it condenses and gives off its heat; therefore, the heat absorbed at the surface of the ocean may moderate climatic conditions far away. For example, heat absorbed in the tropics may be carried to the northern polar regions.

The high heat of fusion of water is another factor serving to reduce radical changes in temperature, since ice is formed from water only after a considerable loss of heat, and it melts only when this amount of heat is replaced. The system:

$$[\text{water}]_{0°\text{C.}} \underset{+\text{heat}}{\overset{-\text{heat}}{\rightleftharpoons}} [\text{ice}]_{0°\text{C}} \qquad (4.1)$$

remains at 0 °C. during the withdrawal of heat, accompanied by ice formation, or the addition of heat, melting the ice.

An additional thermostabilising property of water is its high specific gravity (weight per unit volume) at 4 °C. Because of this, pure water cooled below 4 °C. rises and freezes only at the surface. If it were not for this, bodies of water such as lakes might freeze solid and become uninhabitable.

Finally, water is a very much better heat conductor than any other common liquid. Consequently the heat it absorbs is readily distributed. This tends to equalise the temperature, not only in the external and internal environment but also inside cells.

Water as a Solvent

Besides serving as a thermostabiliser, *water is the best solvent known-more* substances dissolve in it than in any other solvent. Electrolytes dissolved in water readily ionise because water molecules, by virtue of their own charged nature, act as small dipoles attracting the ions of the solute, thereby weakening the attraction between them. In more technical terms it may be said that water is a polar solvent of high *dielectric constant*.

Such a constant can be measured by placing the vapor or solution of a substance between the plates of a condenser and comparing the attraction between the plates with that which occurs in a vacuum. For example, water has a dielectric constant of 178.5, ethyl alcohol 24.3 and benzene 2.28. Ionisation and ionic reactions are of great importance to life, as will be shown in later chapters. For all its importance as a solvent, water is remarkably inert and innocuous and most substances are unchanged by being dissolved in it. Therefore, they may remain in solution in the cell, unaltered until such time as they may be utilised. Water is also necessary for *hydrolyses* and for biological *oxidations,* the two chemical reactions most fundamental to the living cell.

Special Properties of Water

Water has several other properties which make it particularly fit as part of the external and internal environment. It is *transparent to light,* enabling the specialised photosynthetic organelles, the chloroplasts, inside the plant cells to absorb the sunlight-a process upon which depends all life on earth. The transparency of the ocean also enables floating plants and plants attached to rocks to grow below the surface. This property is of considerable importance since marine plants are the ultimate source of food for most marine animals.

The *high surface tension* of water is also an important property since it enables the water to move closer to the surface of the soil than would otherwise be possible, making it available to life. The properties of water and ice continue to be an area of active study. Interest has been focused on the properties of water in the interpretation of resistance of cells to freezing.

Attention has recently been directed to some anomalies in the curve relating to temperature the surface tension and viscosity of water, which shows changes in its slope at 15, 30, 45 and 60° C. Since several properties of biological systems also show changes at these temperatures, such changes in properties of water are thought attributable to the changes in structure in water at these transition points. However, some of the measurements indicating discontinuities in the properties of water are considered by other investigators to be within the limits of error.

Structure of Water

Some of the remarkable properties of water are now interpretable on the basis of its structure. The shape of a water molecule is that of an isosceles triangle; the intermolecular O—H distance is nearly 0.99 A, and the H-O-H angle is not far from 105°. The powerful attraction of the oxygen nucleus tends to draw electrons away from the protons (H nuclei), leaving the region around them with a net positive charge.

Since only two pairs of electrons, around the oxygen atom are shared with the protons, the two other pairs in the eight-electron shell point outward from the O-H bond, attracted to the net positive charge of other water molecules with a force of about 4500 cal./mole. A group of water molecules therefore describes a tetrahedron around the oxygen atom, the positively charged region of one water molecule tending to orient itself toward a negatively charged region of one of its neighbors.

This is shown in Figure elsewhere in this chapter. Each water molecule therefore tends to have four near neighbors and each oxygen atom is the center of a tetrahedron of other oxygens, the 0-0 distance, as determined by x-ray diffraction studies, being 2.76 A.

When ice melts about 15 per cent of the highly regular arrangements found in ice break. Even at higher temperatures, however, some of the structure remains and the high heat of vaporisation of water indicates some quasicrystalline structure even at 100° C.

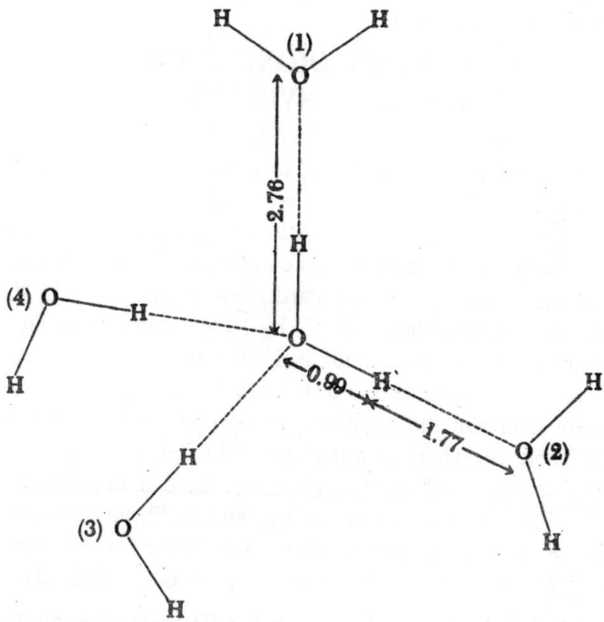

Figure 4.2: The dipolar nature of the water molecule. The diagram shows the orientation of water molecules through tetrahedral coordination in ice. Molecules (1) and (2), as well as the central water molecule, lie entirely in the plane of the paper; molecule (3) lies above this plane, and molecule (4) below it, so that oxygens (1), (2), (3) and (4) lie at the corners of a regular tetrahedron. In water the orientation is not as regular, but multiples occur bound by hydrogen bonds between the molecules (shown in dashes). The distances are given in Angstrom units (10^{-8} cm.).

The type of linkage indicated in Figure elsewhere in this chapter by a dotted line between the proton and the neighboring oxygen atoms is called a *hydrogen bond*. As already pointed out in other chapter, hydrogen bonds are of great importance in aqueous systems and occur in solutions of proteins and other relatively polar substances. Formation and breakage of hydrogen bonds (about 4500 cal./mole) give water most of its peculiar properties. For example, the heat of vaporisation of water is high because all the hydrogen bonds between a water molecule and its neighbors must be broken before it can escape from the surface of the liquid. The high heat of fusion results from similar attraction.

It is difficult to conceive of life without water. For this reason one of the first questions exobiologists ask in considering whether life exists on other planets is: does water occur on the surface or in the atmosphere of the planet?. On the other hand, there is always the question whether life in the dry state or in some other solvent (e.g., ammonia) is truly impossible and inconceivable.

CARBON DIOXIDE IN THE CELL ENVIRONMENT

Carbon dioxide is present in the crust of the earth, in the air and in all natural waters. It is difficult to displace carbon dioxide from water; even after water is boiled, it is rapidly acidified by the reentry of carbon dioxide which dissolves in it. The great solubility of carbon dioxide in water is evident when it is compared with that of other gases present in the atmosphere of the earth. For example, 1.715 liters of carbon dioxide are absorbed at 0° C. by 1 liter of distilled water when the pressure of the carbon dioxide is 1 atmosphere, as compared with 0.049 liters for oxygen and 0.024 liters for nitrogen. Increases in temperature decrease the solubility of carbon dioxide in water, as shown in Table elsewhere in this chapter. Increases in salt content of the water, as measured by the chlorinity, have a like effect. Since carbon dioxide enters water so readily, an equilibrium always exists between the carbon dioxide in water and in air; water can never completely remove it from air, nor can air remove it from the water.

Table 4.1: Solubility or Absorption Coefficients of Some Gases

Temperature	Solubility Coefficients		
°C.	CO_2	O_2	N_2
0	1.713	0.0489	0.0235
5	1.424	0.0428	0.0209
10	1.194	0.0380	0.0186
15	1.019	0.0341	0.0168
20	0.878	0.0310	0.0154
25	0.759	0.0283	0.0143

At room temperature the amount of carbon dioxide dissolved in water is about equal to that in the air with which the water is in contact.

Forming an acid as it dissolves in water, carbon dioxide increases the solvent powers of water for many substances. As a weak acid, it forms one of the primary buffers of the internal fluids of organisms and of terrestrial water.

Table 4.2. Solubility Coefficients of Carbon Dioxide, Showing Variation with Temperature and Chlorinity

Temperature	Chlorinity in gm./Kg.		
°C.	0	16	20
	Solubility Coefficients of CO_2		
0	1.715	1.489	1.438
12	1.118	0.980	0.947
24	0.782	0.695	0.677

The importance of carbon dioxide as a raw material for photosynthesis in plant cells and as one of the wastes from the respiratory processes of plant and animal cells will also be discussed later. It has been calculated that the carbon dioxide in the earth's atmosphere is renewed every 300 to 400 years, that is, each carbon dioxide molecule is incorp-orated into a plant cell every 300 to 400 years.

Carbon dioxide in the air absorbs some of the sunlight entering the atmosphere at wavelengths between 2 and 5 μ, and in this manner tempers the heating of the surface of the earth's crust. It also absorbs strongly the radiations between 13 and 17 μ, near the wavelengths at which the nfrared radiations reradiated from the earth's crust are most intense, thereby delaying the cooling of the surface of the earth with nightfall much as in the "greenhouse effect" described for water vapor.

A current theory ascribes an important role to carbon dioxide in regulating the overall climate of the earth. Thus, the increase in carbon dioxide content of the atmosphere (11 per cent since 1870), from burning of fossil fuels is thought to account for the rising temperature at the surface of the earth for the last 90 years.

If the carbon dioxide concentration in the earth's atmosphere should double, the average temperature increase of the earth's surface would be 3.6° C. If the concentration of carbon dioxide should decrease to one-half of its present value, the average temperature would fall by 3.6° C., enough to cause great ice sheets to form again, as in the great glacial epochs of the earth's history. However, some workers postulate that variations in the output of the sun's radiation may also account for the great changes in climate.

OXYGEN IN THE CELL ENVIRONMENT

Because most cells of organisms are aerobic, requiring oxygen to carry on life activities, oxygen is an important factor of the cellular environment. Oxygen is present in the air and constitutes 20.99 per cent of the atmosphere of the earth. The oxygen in the atmosphere becomes available to living cells as it is dissolved in the liquid medium outside and inside the cell. The oxygen of the atmosphere is continually being replenished as it is liberated during photosynthesis of green cells.

Oxygen is much less soluble in water than is carbon dioxide and diffuses slowly in water. If oxygen were to be distributed by diffusion alone, the oxygen content of a body of water would fall off rapidly below the surface because of the slowness of the diffusion process and aerobic life would be impossible in the depths. However, turbulence caused by currents or welling up mixes a body of water with water containing oxygen and thereby carries oxygen down from the surface waters, greatly facilitating its distribution.

The photosynthetic process of plant cells present in water is also a major source of oxygen and water may even become saturated or supersaturated with oxygen in the neighborhood of photosynthetically active plant cells. On the other hand, if photosynthetic plants are absent and organic material is available, water may become devoid of oxygen as a consequence of the respiration of microorganisms of decay. In muds containing much organic matter, as for example those under brine pools, oxygen is absent within less than a millimeter of the surface. Wet, mucky soil is usually more or less lacking in oxygen, but loose soil contains air pockets with oxygen equal in concentration to that of air.

The solubility of oxygen in water falls with increase in temperature. The deleterious effect of high temperature on animals confined in an aquarium is often the result of lack of oxygen in the water, and care must be exercised to provide an adequate supply.

The solubility of oxygen in water also decreases with increase in salinity. The oxygen content of sea water is therefore slightly less than that of fresh water at the same temperature. The oxygen content of brine, which may contain 10 times as much salt as sea water, is very much less.

The gaseous content of liquids is presented in a variety of ways. In Tables elsewhere in this chapter it is given in terms of the absorption or *solubility coefficient*, which is the volume of gas dissolved per volume of liquid when the partial pressure of the gas in question is 1 atmosphere. It is of greater biological interest to know how much of a particular gas is dissolved in a given volume of liquid in equilibrium with air at 1 atmosphere pressure rather than with the pure gas. According to Dalton's law, the total pressure of a mixture of gases is equal to the sum of the partial pressures of the gases constituting the mixture.

Table 4.3: Solubility Coefficients of Oxygen, Showing Variation with Temperature and Chlorinity.

Temperature	Chlorinity in gm./Kg.			
°C.	0	15	17	20
	Solubility Coefficients of CO_2			
0	.0489	.0406	.0395	.0378
5	.0429	.0359	.0350	.0336
10	.0380	.0321	.0313	.0301
15	.0342	.0292	.0284	.0274
20	.0310	.0267	.0262	.0253

If a liquid is exposed to a mixture of gases, each gas dissolves in proportion to its own *Partial pressure* and independently of the other gases present. The volume of a gas dissolved per unit volume of water is determined by multiplying the solubility coefficient of the gas by the ratio of its partial pressure to the total gas pressure (e.g., at 1 atmosphere).

To obtain the volume of a gas dissolved in any volume of water, the volume of gas dissolved per unit volume of water need only be multiplied by the volume of the water. For example, to determine the oxygen content of 100 ml. of water in equilibrium with air at 20° C and 1 atmosphere pressure, multiply the absorption coefficient, 0.031, by 0.2099, since air contains 20.99 per cent oxygen, and multiply the product by 100. This gives a value of 0.65 ml. oxygen per 100 ml. distilled water.

In physiological literature the volume of the dissolved gas per 100 volumes of liquid is often spoken of as "volumes per cent." However, the concentration of dissolved oxygen, from a physical-chemical standpoint, is better expressed in terms of the partial pressure of oxygen with which the liquid would be in equilibrium, a value which some physiologists call the *oxygen tension* of the liquid.

The partial pressure of the oxygen is readily determined, if the solubility coefficient of the oxygen and the volume dissolved in the liquid under the particular experimental conditions are known, by the use of the equation:

$$\text{volume} = \text{partial pressure} \times \text{solubility coefficient} \qquad (4.2)$$

where the volume of dissolved oxygen is given as volumes of gas per unit volume of liquid, the partial pressure in atmospheres (converted to mm. Hg by multiplying the fraction found by 760 mm.) and the solubility coefficient in volumes of oxygen per volume of liquid under the conditions of the experiment.

The problem of an adequate oxygen supply for all the cells of a multicellular animal is one which in the course of evolution has been solved by the development of oxygen-transporting pigments such as the hemoglobin in our blood. When these are lacking the body fluid contains only as much oxygen as an equal volume of water under the same conditions of temperature and pressure (e.g., in many echinoderms).

However, when an oxygen-transporting pigment is present the amount of oxygen rises in some mollusks to 1 or 2 volumes per cent; in cephalopod mollusks, chironomid larvae (larvae of midges in pond mud) and annelid worms, 2 to 10 volumes per cent; in the blood of fishes, amphibians and reptiles, 5 to 15 volumes per cent; and in the blood of birds and mammals, 15 to 20 volumes per cent. The blood of some diving mammals has an oxygen capacity of 40 volumes per cent; that is, almost twice as much oxygen is combined with the hemoglobin of the blood as is present in an equal volume of air.

The oxygen loosely combined with the oxygen-transporting (respiratory) pigment is carried to all the cells. There a quantity of oxygen is released for use in metabolism in proportion to the difference between the higher oxygen tension in the blood and the lower one in the cells, in accordance with the relationship shown in Figure elsewhere in this chpater.

Some respiratory pigments take up oxygen readily and almost saturate at relatively low oxygen tensions (e.g., the blood of the worm *Arenicola* and the eel), whereas others saturate only at much higher oxygen tensions (e.g., man and the pigeon). The advantage of the blood pigment such as that of *Arenicola is* the ready uptake of oxygen even when the oxygen tension of the environment is relatively low, but such a pigment will not release the bound oxygen until the oxygen tension of the tissue cells is even less than that of the blood (which in this case would have to be below 10 mm. Hg). In the pigeon, on the other hand, an oxygen tension a little below 80 mm. Hg is sufficient to withdraw oxygen from the blood.

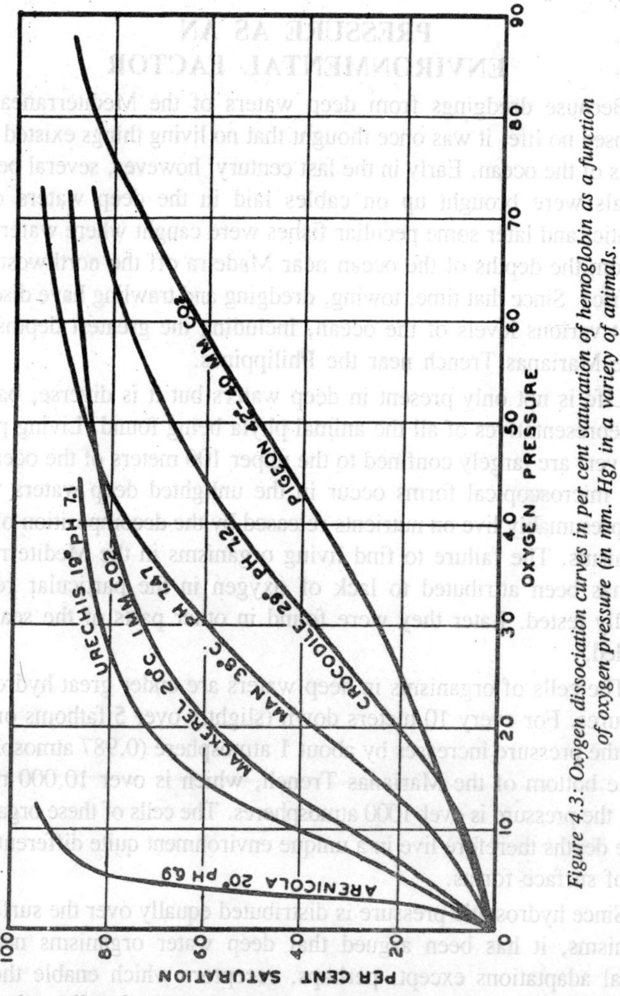

Figure 4.3: Oxygen dissociation curves in per cent saturation of hemoglobin as a function of oxygen pressure (in mm. Hg) in a variety of animals.

In the cells of green plants, the oxygen is supplied during photosynthesis all day. At night the cells must draw upon the oxygen in the atmosphere reaching them by diffusion. According to Rabinowitsch the oxygen in the atmosphere is renewed every 2000 years through photosynthesis. If utilisation of oxygen is faster than can be supplied by diffusion, the plant cells are capable of anaerobic metabolism which can tide them over until oxygen becomes available in daylight. Most organisms, including plants, have a capacity for varying degrees of anaerobic existence and anaerobic metabolism is considered by some as the primeval type of energy liberation.

PRESSURE AS AN ENVIRONMENTAL FACTOR

Because dredgings from deep waters of the Mediterranean Sea disclosed no life, it was once thought that no living things existed in the depths of the ocean. Early in the last century, however, several peculiar animals were brought up on cables laid in the deep waters of the Atlantic, and later some peculiar fishes were caught where water wells up from the depths of the ocean near Madeira off the northwest coast of Africa. Since that time, towing, dredging and trawling have disclosed life at various levels of the ocean, including the greatest depths such as the Marianas Trench near the Philippines.

Life is not only present in deep waters but it is diverse, bacteria and representatives of all the animal phyla being found. Living plants, however, are largely confined to the upper 100 meters of the ocean but some microscopical forms occur in the unlighted deep waters where they presumably live on nutrients released by the decomposition of dead organisms. The failure to find living organisms in the Mediterranean Sea has been attributed to lack of oxygen in the particular regions initially tested. Later they were found in other parts of the sea floor sampled.

The cells of organisms in deep waters are under great hydrostatic pressures. For every 10 meters down (slightly over 5 fathoms or 33.4 feet) the pressure increases by about 1 atmosphere (0.987 atmosphere). At the bottom of the Marianas Trench, which is over 10,000 meters deep, the pressure is over 1000 atmospheres. The cells of these organisms of the depths therefore live in a unique environment quite different from that of surface forms.

Since hydrostatic pressure is distributed equally over the surface of organisms, it has been argued that deep water organisms need no special adaptations except, perhaps, receptors which enable them to perceive each other in the dark. Sensitive, large, stalked or tubular eyes, luminous organs and other receptors are common in deep water fishes and squids.

However, when coastal organisms (mussels, clams, snails, worms and crustaceans) are subjected to pressures gradually increasing to 600 atmospheres, they are first stimulated to greater activity, then depressed and inactivated. Some (starfish and coelenterates) tolerated for one hour pressures up to 1000 atmospheres, which is equivalent to the pressures of the deepest oceans. At this pressure they swelled, but several hours after their return to atmospheric pressure, they recovered. Frogs and fishes were more sensitive and were killed by 300 to 400 atmospheres of pressure. It would seem that organisms of the deep seas have developed some special adaptations enabling them to withstand the great pressures at these depths and still carry on their activities.

The nature of the pressure tolerance adaptations at the cell level have been investigated to only a limited extent and primarily with marine eggs and microorganisms. Sea urchin eggs (*Arbacia punctulata*), beginning to cleave, revert back to the one-celled stage under pressures of more than 400 atmospheres. The contraction and gelation required for cleavage are suppressed. If the pressure is released within a short time, cleavage is resumed. About the same amount of pressure suppresses formation of pseudopodia in *Amoeba proteus;* the plasmagel appears to solate and the viscosity of the cytoplasm falls to a value so low that the food vacuoles and the nucleus sink to the bottom of the cell. It is thus apparent that pressure has a striking effect on sol-gel relationships in the cell.

It might be expected that chemical reactions would also be affected by pressure, and investigations have demonstrated suppression of bacterial luminescence under certain conditions. The luminescence is a result of oxidation of a substrate in the presence of a specific enzyme, luciferase. When pressure is applied, luminescence is gradually quenched until the suspension of bacteria (or a mixture of the enzyme and substrate) becomes dark. If at this point the temperature is raised, luminescence reappears. This has been interpreted as a result of renaturation by heat of the enzyme luciferase denatured by pressure.

Presumably, growth also is suppressed by pressure because cultures of such bacteria as *Bacillus subtilis* and *Pseudomonas fluorescens* fail to grow at pressures of 300 atmospheres or more. Replication of phage in *Escherichia coli is* also prevented by 300 atmospheres of pressure, but as soon as the pressure is released, growth is resumed. Yet there exist marine bacteria, such as *Bacillus submarinus,* isolated from the ocean depths where pressure approximates 500 atmospheres, which grow readily at a pressure of even 600 atmospheres.

Since most of the reactions of the cell are catalysed by enzymes, it is thought that the suppression of cell activities by pressure is a result of action of pressure upon cell enzymes which are proteinaceous in nature. Pressure is found to reduce or prevent the incorporation of precursors into proteins and nucleic acids. Some evidence has already been presented that pressure reverses the denaturing effect of temperature on proteins. It will be recalled that an increase in temperature influences the equilibrium between native and denatured protein in the direction of the denatured form. Denaturation in most proteins is accompanied by an increase in volume as shown by dilatometric measurements.

Since pressure and temperature act in opposition to one another, one would therefore expect pressure to reverse thermal denaturation by suppressing volume increase in accordance with the principle of Le Chatelier, which states that if a system is subjected to a stress it will react in such a way as to reduce the stress. This is indeed found to be the case. Enzymes which are necessary for life activities are catalytic in only a particular configuration, neither fully unfolded nor folded.

Thus, surface bacteria which cannot tolerate the pressures of deep waters at low temperatures are shown to tolerate them and grow if the temperature is increased, as seen in the data of Table elsewhere in this chpater. Perhaps the truly barophilic (literally, pressure-loving) bacteria and other organisms are those in which the enzymes have the proper configuration even under great pressures and at relatively low temperatures.

It is not known by what mechanism pressure causes sex neversal in a copepod or mutation in *Euglena*.

5

WATER ACROSS THE CELL MEMBRANE

Water moves in and out through the plasma membrane at all times, but no effect is demonstrable if the solute concentration is the same on both sides of the membrane. It is when the concentrations of solute are different on the two sides of the plasma membrane that movement becomes apparent by shrinkage or swelling of the cell. Water moves in or out of the cell until the concentration of solute is equalised between the inside and the outside of the cell. *Osmosis* is the diffusion of water (or more generally, a solvent) through a membrane in response to a concentration gradient.

The principles governing osmotic transfers have been studied with artificial, natural and living cell membranes.

PASSAGE OF WATER ACROSS ARTIFICIAL MEMBRANES

Many artificial membranes have been devised as analogues to the plasma membrane to facilitate study of osmotic transfers. A membrane of copper ferrocyanide, formed by the interaction of potassium ferrocyanide and copper sulfate, has been a favorite artificial membrane since it was first introduced by Traube in 1867. When a crystal of copper sulfate in a piece of gauze is suspended in a test tube filled with a solution of potassium ferrocyanide, the copper sulfate dissolves and interacts with the ferrocyanide, forming a membrane. Water enters this membrane causing the bag or "cell" to swell and burst. A new membrane precipitates in a new position as more of the copper sulfate is dissolved and additional water is taken up, and the process repeats itself until the copper sulfate is completely dissolved. A similar membrane for experimental study was devised by Pfeffer in 1877. Potassium ferrocyanide solution is"placed inside an unglazed pottery jar surrounded by a solution of copper sulfate.

An electric current is then passed between the two solutions. The membrane of copper ferrocyanide which is precipitated in the small pores of the unglazed jar allows only water to pass through it. If the jar is then filled with 1 molal NaCl and is suspended in a beaker of distilled water, water molecules pass through the membrane but salts

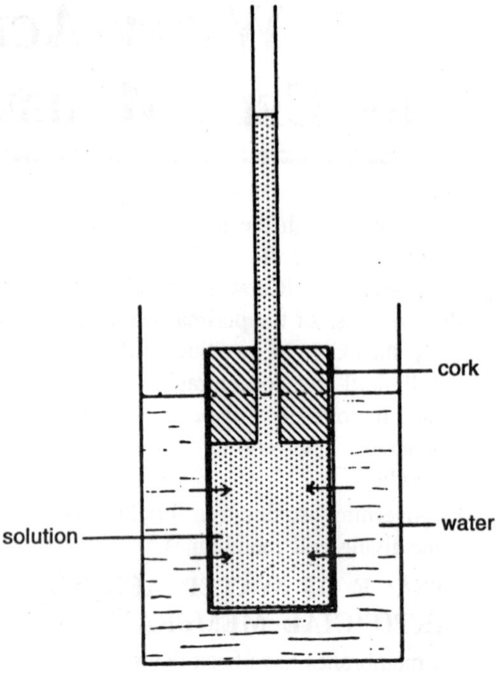

Figure 5.1: A simple osmometer. The pores in the jar have been coated with a precipitate of copper ferrocyanide, which is permeable to water but not to solutes.

cannot, and since the water molecules cannot get out as readily as they enter, because they collide with sodium and chloride ions inside the jar, water accumulates inside the jar. If a tube is fitted to the mouth of the jar, water rises in the tube until the pressure developed by the rise of water equals the *osmotic pressure* (the tendency of the water molecules to enter the jar in response to the concentration gradient between the inside and the outside of the jar). Such a membrane, which is permeable to solvent only-in this case, water is said to be *semipermeable*. In aqueous systems a semipermeable membrane implies a membrane permeable to water only.

A better way of measuring the osmotic pressure than by the height of water in a tube is by means of the mercury osmometer introduced by Pfeffer and illustrated in Figure elsewhere in this chpater. An *osmometer* is a device used to measure osmotic pressure. It must have a semipermeable membrane which allows water (solvent) but not the solute to pass. The physical chemist employs still other devices to determine osmotic pressure. One such method is measuring the pressure that just prevents the entry of water (solvent) through the semipermeable membrane; this is taken as a measure of the osmotic pressure. Another method is the Hepp osmometer in which very small quantities of material can be used and a direct reading of the osmotic pressure may be made. Or, alternatively, the osmotic pressure registered on a diaphragm may be transduced to an electrical change readily measured and recorded.

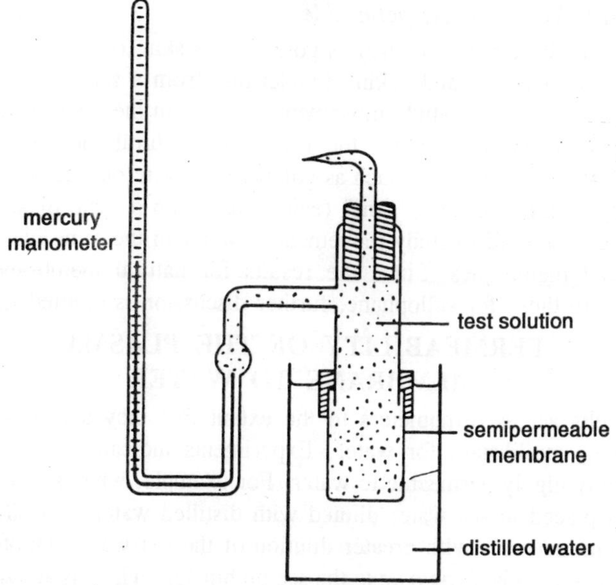

Figure 5.2: Pfeffer's osmometer with mercury manometer to measure the osmotic pressure. The test solution is kept inside the jar, in the walls of which has been precipitated a copper ferrocyanide membrane. Entry of water occurs until the osmotic pressure is counterbalanced by the pressure of the column of mercury.

An artificial membrane often used in biological studies is cellophane dialysis tubing. If such a membrane is fastened over a thistle tube and the tube is filled with concentrated sodium chloride solution and immersed in distilled water, water will enter and the volume of solution in the thistle tube will increase as a consequence, indicating that cellophane is permeable to water, Tests for chloride in the distilled water outside, however, show that sodium chloride is also coming through the cellophane.

But the rise in the level of the solution in the tube proves that entry of water is greater than outward movement of salt. If glucose is used instead of sodium chloride similar results are obtained, except that the outward movement of glucose is slower than that of salt. To show that passage of solute is not the result of a mechanical leak, Congo red, a colloidal dye, is added to the solution in the thistle tube during such an experiment. It will be recalled that colloids do not pass through cellophane; therefore, the dye will be retained by an intact membrane, but the appearance of a pink colour in the water surrounding the tube will indicate a tear in the membrane. A membrane which lets one material (water) pass through more readily than another (salt or sugar) is said to be *selectively permeable.*

Natural membranes, such as goldbeater's skin (outside membrane of the ox intestine) and "skin" (epidermis) from a scale of an onion bulb, can be used to study movement of water in the same manner as artificial membranes. Since the results for natural membranes are simi-Natural membranes, such as goldbeater's skin (outside membrane of the ox intestine) and "skin" (epidermis) from a scale of an onion bulb, can be used to study movement of water in the same manner as artificial membranes. Since the results for natural membranes are similar to those for cellophane, further discussion is omitted here.

PERMEABILITY OF THE PLASMA MEMBRANE TO WATER

Cells act as osmometers to the extent that they act as systems attaining equilibrium for water. Experiments indicate that cells are generally highly permeable to water. For example, when a sea urchin egg is placed in sea water diluted with distilled water it swells. The swelling increases with greater dilution of the sea water. Diluted sea water is said to be *hypotonic* to the sea urchin egg, which is accustomed to undiluted sea water. Swelling continues with dilution until the membrane bursts and cytolysis occurs.

When the egg is placed in sea water concentrated by evaporation, it shrinks, and the shrinkage increases as the medium becomes more concentrated. Such a medium is said to be *hypertonic* to the sea urchin egg. The egg in unaltered sea water does not swell or shrink, and the sea water is said to be *isotonic.*

Experiments with red blood cells (erythrocytes) indicate that they also are permeable to water. If the erythrocytes are placed in a hypertonic solution they shrink in a characteristic way and are said to be crenated. When placed in a hypotonic salt solution they swell and may become spherical instead of discoidal. Erythrocytes become hemolysed and lose their hemoglobin when placed in excessively dilute salt solutions.

A concentration of sodium chloride of approximately 0.7 per cent is isotonic to a red blood cell of a frog, and.a concentration of 0.9 per cent is isotonic to mammalian erythrocytes. Many of the problems of water permeability in animal cells are discussed by Dick. Plant cells, too, are quite permeable to water. The cells of a fresh water plant like *Elodea* are normally turgid; that is, the protoplast is held firmly against the cell wall, and the entire leaf has a certain rigidity as a result of the collective rigidity of the individual cells. If the plant cells are placed in a concentrated salt solution, they become plasmolysed, and the protoplast of each cell contracts about the shrunken vacuole.

If the hypertonic salt solution is replaced by fresh water, each cell regains water and the protoplast again fills the space within the cell wall. When *Elodea* is placed in a series of solutions containing increasing concentrations of salt, plasmolysis occurs in its cells when the concentration of a salt, such as potassium nitrate, reaches about 0.6 per cent. Less salt than this is present in the fresh water of ponds and streams in which *Elodea* is found. The rigid cell wall, limiting the volume, prevents excessive uptake of water. However, if the cell wall is first digested by cellulase then the naked protoplast will swell with uptake of water.

Cells plasmolysed by hypertonic solutions return in time to their initial volume (deplasmolysis). Deplasmolysis is rapid in solutions of solutes which penetrate rapidly, and it is slow in solutions of solutes which penetrate slowly. For example, the initial shrinkage of an egg or of *Elodea* cells, when in hypertonic solutions, indicates that water leaves the cell more readily than the solute molecules enter, but subsequent swelling shows that the solute molecules have penetrated the cells. Solute molecules accumulate in the cell and water reenters, causing the cell to swell.

The membrane of a cell is therefore not semipermeable but selectively permeable; that is, it is more permeable to water than to other substances. This is found to be true of all living cells studied. It would appear that the plasma membrane is freely permeable to water and permits water to enter when the cell is placed in hypotonic solutions and permits water to leave when the cell is placed in hypertonic solutions.

Bacteria, like plant cells, are also held within cell walls and do not appear to swell in hypotonic solutions. However, if the cell wall of the bacteria is removed by lysosyme, the protoplast swells and shrinks like any other cell.

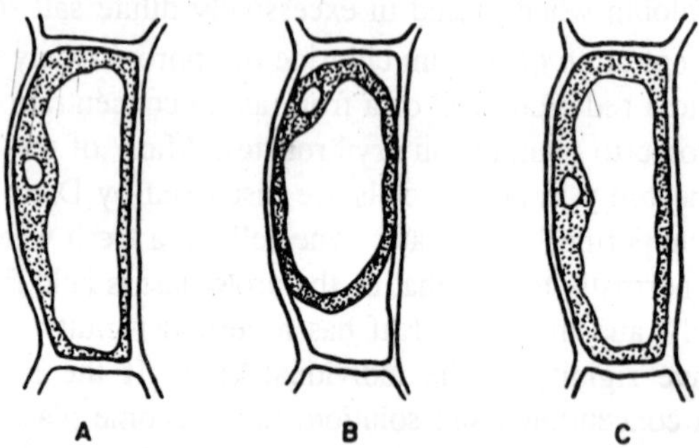

Figure 5.3: A plant cell (A) undergoes plasmolysis (B) when it is placed in a hypertonic solution. Deplasmolysis (C) occurs after a period of time, but only if the solute can penetrate. Time for deplasmolysis depends upon the rate of entry of the solute.

The cell organelles, as we have seen, are covered by unit membranes, which appear to be selectively permeable like the plasma membrane. Thus mitochondria shrink or swell in response to the concentration of solute in the bathing media. Less is known of the membranes of the endoplasmic reticulum, Golgi complex and other organelles because they are less susceptible to experimental study in this regard. However, it seems likely that these membranes are also freely permeable to water. Their capabilities in concentration of certain ions at the expense of metabolic energy is dealt with elsewhere.

OSMOTIC PRESSURE OF NONELECTROLYTES

As already discussed, when a solution of a nonelectrolyte in a thistle tube is separated from distilled water by a semipermeable membrane, water will pass through the membrane into the solution causing an increase in volume and a rise of solution in the tube. The pressure exerted by the solution in the tube is proportional to the height (h) of the column and is called the osmotic pressure (π). Water will continue to increase the volume of solution until the pressure of the column on the membrane just prevents the entry of water into the membrane. At that time equilibrium will have been established, and

$$\pi = h \times k \qquad (5.1)$$

where k is a constant of proportionality for a given temperatureand concentration. The higher the concentration of solute, the higher the rise of the water will be.

The plant physiologist, Pfeffer, made measurements of osmotic pressure of various solutes by using a copper ferrocyanide osmometer and found that the osmotic pressure divided by the concentration of solute is constant, but he did not realise the full significance of his data. The physical chemist, van't Hoff, showed that Pfeffer's data could be taken as evidence of the applicability of the gas laws to osmotic pressure. A gram molecular weight of a gas at atmospheric pressure occupies 22.4 liters of space.

If it is compressed into the confines of a liter it exerts 22.4 atmospheres pressure. If a solute is compared to a gas, then a gram molecular weight of the solute in a liter of solution, if it acts like a gas, should exert a pressure of 22.4 atmospheres. When analysed, Pfeffer's data showed that a gram molecular weight of a nonelectrolyte dissolved in a liter of water exerts an osmotic pressure of approximately 22.4 atmospheres, confirming van't Hoff's hypothesis. The agreement is only approximate, since exact agreement would occur only if the molecules of solute occupied no volume and the attractions between them were nil.

Table 5.1: Pfeffer's Data for Osmotic Pressure at Different Concentrations of Sucrose

Concentration (C) in percent	Osmotic Pressure (π) in Atmospheres	$\frac{\pi}{C}$
1	0.70	0.7
2	1.34	0.67
4	2.74	0.88
8	4.10	0.88

The gas law may be written PV = nRT, where P is the pressure in atmospheres, V the volume in liters, n the number of moles, R the gas constant (equal to 0.082 liter atmospheres per degree per mole) and T the absolute temperature. By rearrangement,

$$P = \frac{n}{V} \times RT \tag{5.2}$$

Since the number of moles of solute divided by the number of liters of solvent in which it is dissolved is equal to the molal concentration (moles in a kilogram of solvent), C may be substituted for n/V. Then,

$$P = CRT \quad (5.3)$$

For osmotic pressure this is usually written:

$$\pi = CRT \quad (5.4)$$

in which π denotes the osmotic pressure. This relationship can be derived from basic thermodynamic principles.

For a gram molecular weight of osmotically active ideal nonelectrolyte, C is 1, R is 0.082 liter atmospheres per degree per mole, and the absolute temperature at 0° C. is 273°. Then

$$\pi = (1 \text{ mole/liter}) (0.082 \text{ liter atm./degree/mole}) (273°). \quad (5.5)$$

or g = 22.4 atmospheres

Since the gas laws were developed for perfect gases whose molecules have no volume and do not attract one another, their application to solutions is even more of an approximation than their application to gases. The gas law can be applied only to weak solutions at best. Deviations become marked at higher concentrations.

OSMOTIC PRESSURE OF ELECTROLYTES

Since electrolytes dissociate, each molecule gives rise to a number of ions. Because the osmotic pressure is determined by the number of particles present in the solution, for a given *molal* concentration an electrolyte will exert a greater osmotic pressure than a nonelectrolyte. The magnitude of the difference depends on the degree of dissociation and the number of ions produced.

Therefore, the osmotic pressure of an electrolyte, calculated by the gas law, must be multiplied by a factor correcting for dissociation. The correction factor is less than the total number of ions formed from a molecule of an electrolyte, because the activity of the ions, rather than their concentration, determines the osmotic pressure. Ionic interference reduces the effective concentration of the ions; e.g., unlike ions attract one another, reducing the number of ions free to act. The activity of a given electrolyte varies at different concentrations, being generally greater at low concentrations in which interference between ions is least.

The determination of the activity for each concentration of an electrolyte is most easily done by cryoscopy (freezing point determination), which measures the number of particles active, not just their concentration. Then, by comparing the freezing point depression of an electrolyte with that of an ideal nonelectrolyte of the same concentration (usually 1 molal of each is used) a factor called the *cryoscopic coefficient,* G, may be obtained:

$$G = \frac{\Delta T_{fp}}{\Delta T_{\overline{fp}}} \tag{5.6}$$

where ΔT_{fp} is the freezing point depression produced by a given molal concentration of the electrolyte in question and ΔT_{fp} is the lowering of the freezing point by the same molal concentration of an ideal nonelectrolyte. A few cryoscopic coefficients are given in Table elsewhere in this chapter. The osmotic pressure of an electrolyte can be calculated from the gas law equation (5.4), but the concentration term in the gas law equation must be multiplied by the cryoscopic coefficient, G, for each concentration of electrolyte

$$\pi = (\text{C} \times G)RT \qquad 5.7)$$

since G measures the relative increase of number of particles by ionisation and includes an activity correction.

Table 5.2: Cryoscopic Coefficients (G) at Various Molal Concentrations

Electrolyte	0.02	0.05	0.1	0.2	0.5
$MgCl_2$	2.708	2.677	2.658	2.679	2.896
$MgSO_4$	1.393	1.302	1.212	1.125	—
$CaCl_2$	2.673	2.630	2.601	2.573	2.680
LiCl	1.928	1.912	1.895	1.884	1.927
NaCl	1.921	—	1.872	1.843	—
KCl	1.919	1.885	1.857	1.827	1.784
KNO_3	1.904	1.847	1.784	1.698	1.551

COLLIGATIVE PROPERTIES AND THE DETERMINATION OF OSMOTIC PRESSURE

The methods already discussed for the direct measurement of osmotic pressure are not very suitable for the measurement of the osmotic pressure of cell contents because the materials are available only in relatively small quantities. Therefore, use has to be made of some properties of solutions that are functions of the same variable, namely, the *number* of particles in solution and not the *kind* of particles.

Those properties of a solution that depend upon the number of particles present per unit volume of solution are called its *colligative properties,* and they are a manifestation of a phenomenon common to all of them; i.e., solute molecules decrease the tendency of water molecules to move from one solution to another or from one phase to another. The *colligative properties* of solutions are (1) the osmotic pressure, (2) the freezing point depression, (3) the boiling point elevation and (4) the vapor pressure.

The dependence of osmotic pressure upon the number of active particles has already been discussed. The greater the number of particles, assuming that they are free to move, the greater will be their interference with movement of water molecules from a solution through a membrane. This interference with movement of water molecules also affects the freezing point, boiling point and vapor pressure in a like manner. The freezing point is lowered because solute molecules interfere with the formation of ice crystals.

The boiling point is raised because the solute particles in solution interfere with the establishment of an equilibrium between the water molecules and the atmospheric pressure. Similarly, the vapor pressure, which depends upon the evaporation of water from the solution, is decreased by the interference of the solute particles with the escape of water molecules, and the decrease is proportional to the concentration of solute. It is self-evident, therefore, that if one of the colligative properties is known, any of the others can be calculated, since they are all related through the same variable.

The vapor pressure determination is perhaps the simplest, as seen from the following example. When two beakers containing aqueous solutions of different concentrations of solute are placed in an airtight enclosure, the water molecules leave and reenter both solutions continually, but because exit of water molecules is less impeded, in the dilute solution, more water will eave it than leaves the more concentrated solution.

Thus, the dilute solution loses more water than it receives and the concentrated solution receives more water than it loses until the two become equal in concentration. Since the temperature falls in the solution losing water, and since the change in temperature is proportional to the difference in concentrations of the two solutions, the vapor pressure can be calculated from the change in temperature at equilibrium. Using this principle, an elegant method for determining osmotic pressures of very small samples of fluid has been developed. This method employs thermocouples to measure the temperature, because thermometers are not sensitive enough for such small samples of fluid. A thermocouple

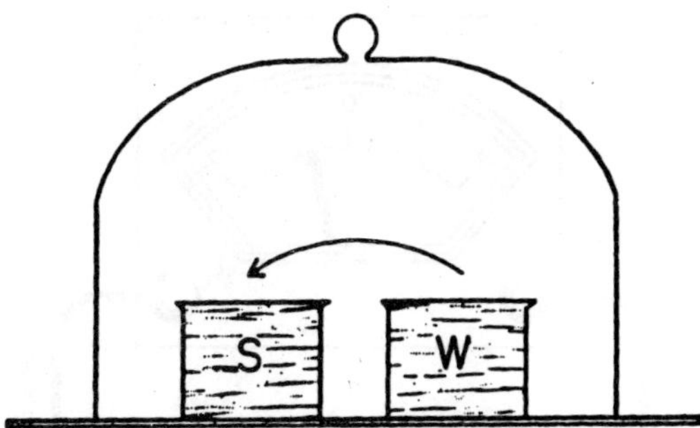

Figure 5.4: Water distils over into a solution because of a difference in vapor tension between pure water (W) and the solution (S). The solute molecules reduce the escaping tendency of the water molecules in the solution. The temperature of distilled water is lowered by evaporation from its surface. This lowering of temperature can be measured by a thermocouple.

consists of a junction of two metals in which a change in temperature induces an electromotive force creating a current measurable with a galvanometer. First the junctions are calibrated when one is placed in a solution of known concentration (and osmotic pressure) and the other in distilled water. A series of solutions of known osmotic pressures are tested in this manner, and a graph is constructed that relates galvanometric deflection to osmotic pressure. The galvanometric deflection produced by a sample of unknown osmotic pressure is then determined in the same way and under the same conditions. This galvanometric deflection, placed on the standard curve, at once enables one to determine the osmotic pressure of the sample. Commercially produced instruments for vapor pressure measurements are now available.

Another method for determining osmotic pressure from vapor pressure is Barger's method The fluid to be tested is taken up into one end of each of a series of capillary tubes. On the other end of each tube, a sample from a series of known concentrations is introduced, an air space remaining between the two solutions in the tube. The initial length of each solution in each tube is marked.

The unknown solution is tested in this way against the series of solutions of known concentr-ations. Each tube is examined periodically under a microscope to see if the test sample has gained or lost water. This is continued until it is definitely established that a pair is found in which no water exchange occurs. If no water is exchanged between the two samples, the vapor pressures in the two solutions are equal. Then the osmotic pressures also are equal in the two, and that of the unknown can be calculated in terms of the known. This method has the advantage of requiring a minimum of equipment, but it is somewhat

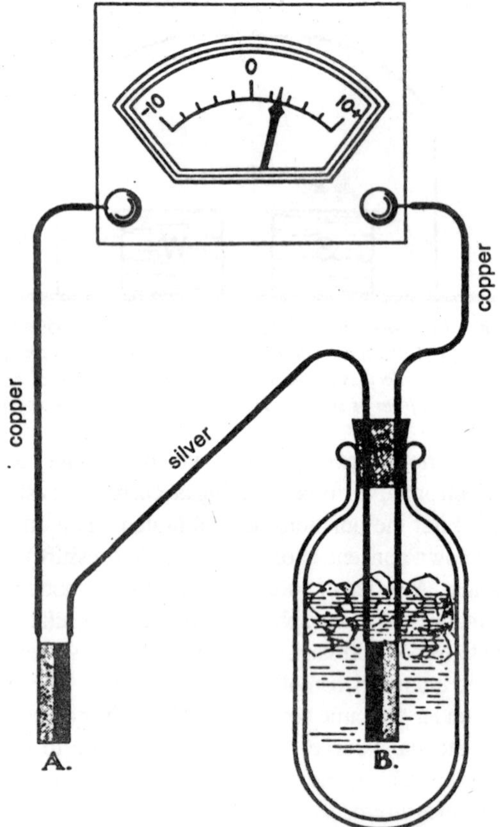

Figure 5.5: An elementary form of thermocouple for measuring change in temperature. Junction B is always kept at some convenient known temperature, e.g., on ice.

tedious in operation. In biological work, determination of the freezing point of a solution has been used most extensively for determining the osmotic pressure, because the equipment is generally available and the measurement can be made without altering the nature of the solution.

It is known that the freezing point of an aqueous solution containing a gram molecular weight (1 molal) of ideal nonelectrolyte is depressed by 1.86° C. The solution is placed in a freezing mixture with a thermometer (or thermistor, in the Fiske osmometer), and the temperature is taken at the time ice crystals appear. The temperature is taken again when the crystals disappear on warming. From the depression of the freezing point the osmotic pressure is readily calculated with the following equation:

$$\frac{\Delta_x}{\Delta_m} = \frac{x}{22.4 \text{ atm.}} \tag{5.8}$$

$$x = \frac{\Delta_x}{\Delta_m} \times 22.4 \text{ atm.} \tag{5.9}$$

$$\text{or } x = \frac{\Delta_x \cdot 22.4}{1.86} = 12.06\ \Delta_x \tag{5.10}$$

Here Δ_x = the depression of the freezing point produced by the "unknown" solution whose osmotic pressure is sought. Δ_m = the depression of the freezing point produced by a 1 molal solution of ideal nonelectrolyte and x = the osmotic pressure of the "unknown" solution. At present methods are available for freezing point determinations by this method on micro-quantities of fluid.

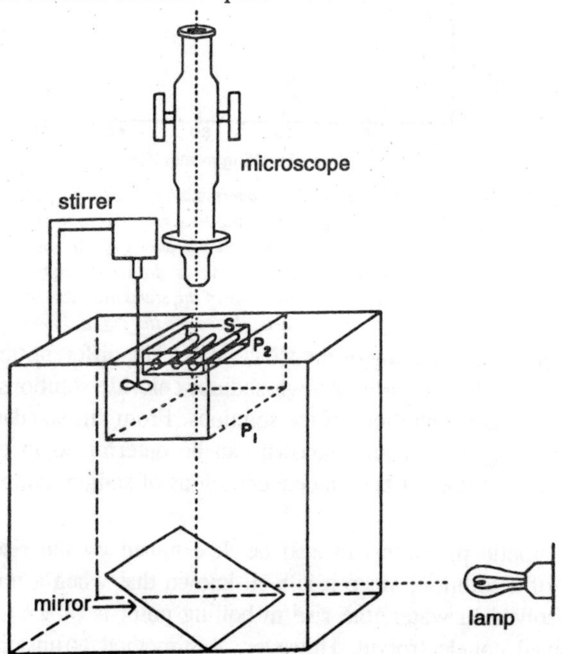

Figure 5.6: Determining the melting point of minute quantities of biological materials.

A variant of the freezing point method may be used for very small quantities of fluid. Test solutions are drawn up into capillary tubes and frozen. Their melting is then observed through crossed polaroids, under a microscope. Standards of known concentrations of sodium chloride (or other solute) are prepared in the same manner so that standard and test solutions can be examined simultaneously for comparison of the times

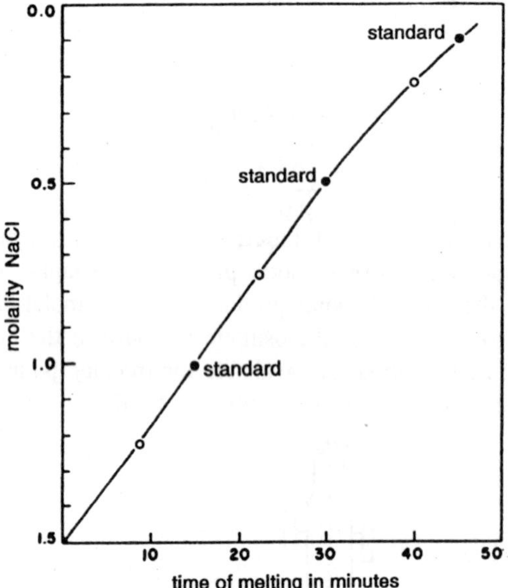

Figure 5.7: Use of the melting method for determination of osmotic pressure. A standard curve is first established by plotting data for the time required to melt frozen solutions of known concentrations of NaCl against the concentrations of the solutions. Three such points are shown as black dots, and the curve is drawn through these points. The concentration of an unknown solution, in terms of the standards, can be determined if the melting time (open circle) for a sample is placed on the curve.

required for the melting of the ice crystals. The different times required for melting the ice in the known sodium chloride solutions are plotted against the concentrations of the solutions. From this standard graph the osmotic pressure of an unknown can be determined in terms of the osmotic pressures of known concentrations of sodium chloride or other solute.

Osmotic pressure can also be determined by the rise in boiling point of a solution, since it is well known that when a nonelectrolyte is dissolved in water, the rise in boiling point is 0.52° C. per molal dissolved nonelectrolyte. However, this method is unsatisfactory for determining the osmotic pressures of biological solutions, because proteins coagulate when they are boiled.

OSMOLAL SCALE FOR BIOLOGICAL STUDIES

It is the concentration of osmotically active particles inside cells or in the fluids bathing the cells that is of consequence to movement of water between the medium and the cell, not whether the particles are ionic or nonionic, organic or inorganic. It is, therefore, useful to have terminology in terms of osmotic activity. For example, the *osmole* is a gram molecular weight of osmotically active particles in 1000 gm. of solvent (water, in biological systems). An osmole exerts 22.4 atm. of osmotic pressure and an osmolal solution lowers the freezing point 1.86° C. Frequently osmolal concentrations are given in milliosmolal units (a milliosmole is 1/1000 osmole) because of the low concentration of solutes in biological materials and media. The osmolal concentration of a solution is determined by any one of the colligative properties.

SOURCE OF ENERGY FOR WATER MOVEMENT

Traversing the plasma membrane by any molecule requires energy. Molecules at a temperature consistent with life processes possess kinetic energy. Some molecules in a given population possess more than the average amount of energy. Such molecules with at least the required energy of activation can surmount the hurdles required for their entry by diffusion. When water moves against a gradient metabolic energy is required.

CELLS AS OSMOMETERS

The living cell, having a selectively permeable membrane rather than a semipermeable membrane, could hardly be expected to behave like a perfect osmometer. However, a cell might.perhaps show some approximation to an osmometer if in response to a concentration gradient water were to pass through the membrane much more readily than would the solute.

Table 5.3: Arbacia Egg as an Osmometer.

Relative Pressure	Volume Observed	πV	π(V – b)
1.0	2,121	2,121	1,881
0.9	2,316	2,084	1,868
0.8	2,570	2,056	1,864
0.7	2,922	2,045	1,878
0.6	3,420	2,053	1,909
0.5	4,002	2,002	1,881

However, one correction must be made before such a test can be applied with fairness. The cell contains in its structures much protein and lipid, which add to the total volume of the cell, thereby excluding from that volume a corresponding volume of solution. The volume occupied by cell structures and called the *nonsolvent volume* (usually designated b), must be subtracted from the term for total volume in the equation for osmotic pressure: $\pi V = nRT$. The equation therefore becomes:

$$\pi(V - b) = nRT \tag{5.11}$$

where b represents the nonsolvent volume and the other terms have the meanings previously assigned. In Table 12.3 the nonsolvent volume is given as 12.5 per cent of the total volume of the cell, or 0.125. When correction for the nonsolvent volume is made, the sea urchin egg in diluted sea water is observed to behave reasonably well as an osmometer, the values in the fourth column being more nearly alike than those in the third column, where no correction was made. In other cells, where solutes pass through the membrane more rapidly than they do in the sea urchin egg, the cells do not act even approximately like osmometers.

To compare the relative permeabilities of different kinds of cells to water, it is necessary to obtain the volume of water entering per unit area of cell surface per unit time per unit difference in osmotic pressure between the inside and the outside of the cell. A permeability constant, k, for water is given in these terms by the following equation:

$$\frac{dV}{dt} = k\,A(\pi_{in} - \pi_{out}) \tag{5.12}$$

where V is the volume of the cell, dV, a small change in volume; t is the time, dt, a small change in time, A is the area of the cell surface; π_{in} is the osmotic pressure of the cell; and π_{out} is the osmotic pressure of the external medium.

The low permeability rates to water of different cells, as determined primarily by swelling of cells in hypotonic media, are given in Table elsewhere in this chapter. Almost a two hundred fold variation in permeability to water is observed in these cells. Cells with a higher surface to volume ratio have a higher permeability to water than those with a lower ratio. It is perhaps significant that the fresh water forms, in which the greatest gradient is found between the inside and the outside of the cell, have the lowest permeability constants for entry of water. Protozoan cells in fresh water are forced to void the constant excess of incoming water by means of contractile vacuoles. A low permeability to water reduces this work to a considerable degree.

Water movement into and out of a cell is not always as simple as this discussion would lead one to believe. It is important to point out that the permeability constant of a cell to water is only an approximation. For example, difference in osmotic pressure cannot be the only factor

Table 5.4: Comparison of Permeabilities of Some Cells in Water.

Species	Permeability Constant
Amoeba proteus	0.026-0.031
Pelomyxa carolinensis	0.023
Fresh water peritrichs	0.125-0.25
Arbacia egg	0.4
Human erythrocyte	3.0

governing entry or exit of water, since in some cells, even the inward and outward permeability to water is different, indicating operation of some biological factor as well. Furthermore, depending upon its physiological state, the same cell may show changes in its permeability to water. For example, the permeability of some echinoderm eggs to water doubles, triples or quadruples on fertilisation of the eggs. After cell division occurs, permeability to water may fall back to the original value. It is curious that some cells are much more permeable to water than others and some cells even secrete water; for example, the gland cells in the carnivorous bladder plant *Utricularia,* set the trap by this means. All of these are interesting and fundamental problems involving the movement of water across the cell membrane.

6

Energy of Life

Every activity of a living cell or organism requires energy. Movement, synthesis of needed compounds, manufacture of new cells, and transport of materials are just a few of the life processes that require energy. Because an organism has no way of creating new energy or of recycling the energy that it has used, life depends upon a continuous input of energy. This means that there is a one-way flow of energy through any individual organism and through the ecosphere. Producers trap light energy from the sun during photosynthesis and incorporate some of that energy in the chemical bonds of carbohydrates and other organic compounds. Then, a portion of that chemical energy can be transferred to the consumers that eat the producers and to the decomposers that feed on all of them, sooner or later.

An organism must be considered an open system with respect to energy because of the one-way energy flow through it. Energy is captured, temporarily stored, and then used to perform biological work. During these processes it is converted to heat and dispersed into the environment. Because this energy cannot be reused, organisms must continuously obtain fresh supplies of it. But just how do they do so? In this chapter we will focus on some principles of energy capture, storage, transfer, and use. In the following two chapters we will explore some of the main metabolic pathways used by cells in their continuous quest for energy.

WHAT IS ENERGY?

Energy may be defined as the ability to cause various types of changes, or more simply, as the capacity to do work. Here we are concerned with the ability of living systems to do biological work. It takes the same amount of energy to light a 75-watt bulb as to keep your brain in operation while you read these words. And at this very moment you are expending considerable amounts of energy to maintain your breathing, concentrate urine in your kidneys, digest food, circulate your blood, and maintain countless other metabolic activities.

These are all forms of biological work. Energy can exist in several forms. These include heat, electrical, mechanical, chemical, sound, and radiant energy (the energy of electromagnetic waves, such as radio waves, visible light, x-rays, and gamma rays).

Potential and Kinetic Energy

Energy can be described as potential or kinetic. *Potential energy* is stored energy; it has the capacity to do work owing to its position or state.

In contrast, *kinetic energy* is the energy of motion. An example of the conversion of potential energy to kinetic energy is the release of a

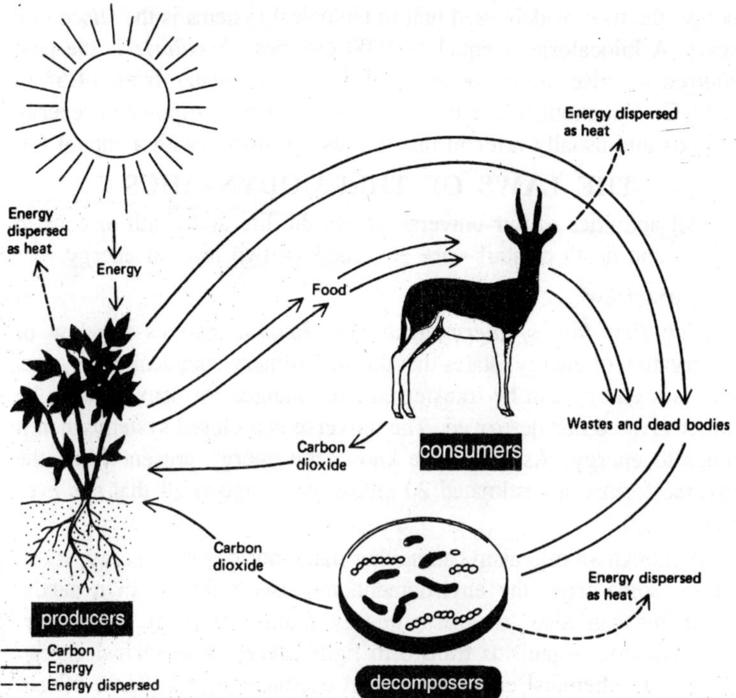

Figure 6.1: Flow of energy through the ecosphere. Carbon and many other chemical elements are continuously recycled. Energy, on the other hand, cannot be recycled.

drawn bow: The tension in the bow and string represents stored energy; when the string is released, this potential energy is released so that the motion of the string propels the arrow.

It would require the input of additional energy to draw the bow once again and restore the potential energy. Most of the actions of a complex organism involve a complex series of energy transformations. For example, to prepare for a running event, athletes eat foods that build up their reserves of glycogen. During the event an athlete's body continuously converts the energy stored in the glycogen into the kinetic energy used to run the race.

Measuring Energy

To study energy transformations, scientists must be able to measure energy. How is this done? Heat is a convenient form in which energy can be measured because all other forms of energy can be converted into heat. In fact, the study of energy has been named *thermodynamics*, that is, heat dynamics. Although several units may be used in measuring energy, the most widely used unit in biological systems is the *kilocalorie* (kcal). A kilocalorie is equal to 1000 calories. A *calorie* is the heat required to raise the temperature of 1 gram of water from 14.5° to 15.5° C. Nutritionists use the kcal in measuring the potential energy of foods and usually refer to this unit as a Calorie (with a capital C).

THE LAWS OF THERMODYNAMICS

All activities of our universe—from the life and death of cells to the life and death of stars—are governed by two laws of energy.

The First Law

The *first law of thermodynamics*, known also as the law of conservation of energy, states that during ordinary chemical or physical processes energy can be transferred and changed in form, but can be neither created nor destroyed. The universe is a closed system when it comes to energy. As far as we know, the energy present when the universe formed an estimated 20 billion years ago is all that can ever exist.

Although an organism can neither make nor destroy energy, it can capture some from the environment and use it for its own needs. Organisms can also transform energy from one form to another. Photosynthetic organisms transform light energy to electrical energy and then to chemical energy stored in chemical bonds. Some of that chemical energy may later be transformed by an animal that eats the plant or alga to the mechanical energy of muscle contraction or to

1. The polarised charges inside and outside the membrane of the resting muscle cell represent potential electrical energy. When a neuron releases a chemical that stimulates depolarisation of the charges, this in turn stimulates the muscle to *initiate* contraction.
2. Chemical energy is stored in the high energy phosphate bonds of a molecule called ATP. Breaking these bonds releases energy to carry out muscle contraction.
3. This energy released inside each muscle cell causes submicroscopic thick filaments to pull on thin filaments by means of cross bridges. The thin filaments move past the thick ones, causing the cell to shorten.

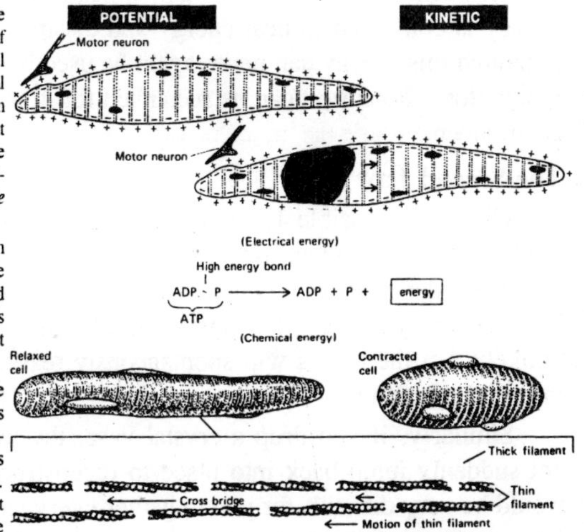

4. The archer's muscles contract, pulling on the bones, which then move and transmit the muscular force to the bow.
5. The archer has drawn the bow. The energy stored in the drawn bow is potential mechanical energy. When the archer releases her fingers the string moves forward and the arrow is propelled toward a target.

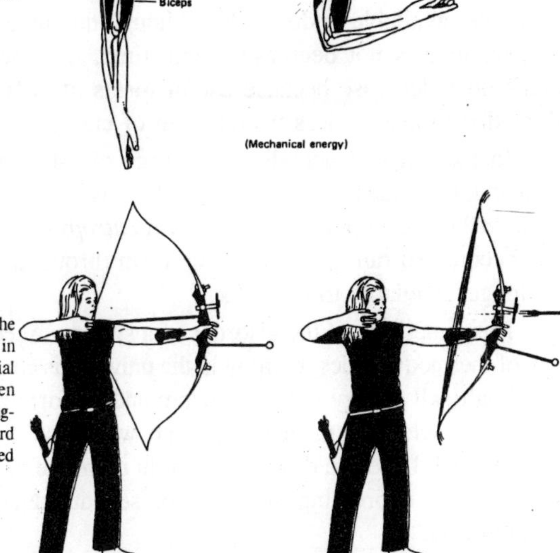

Figure 6.2: Potential energy is stored energy. Kinetic energy is the energy of motion. Shooting a bow and arrow illustrates some of the energy transformations that take place in living systems generally.

another needed form. As these many transformations occur, some of the energy is converted to heat energy and dissipated to the environment. Although this energy can never again be used by the organism, it is not really "lost" because it can be accounted for in the surrounding physical environment.

The Second Law

Given all possible forms of arrangement, order is an extremely unlikely state. From your own experiences you might have observed that creating order requires an input of energy or work. For example, if you spill the many pieces of a jigsaw puzzle onto the floor, it is highly unlikely that the pieces will spontaneously reassemble themselves into the original picture.

Similarly, if you drop a crystal vase, the fragments of glass will not suddenly jump back into place to reconstruct the vase. In fact, it would be very difficult for you to gather every small shard of glass and fit these fragments together to mend the vase. Out of the multitude of possible ways in which the pieces can be assembled, only one represents the highly ordered form that is the vase or the finished puzzle.

The **second law of thermodynamics** states that disorder in the universe is continuously increasing. This law explains that in any process requiring or releasing energy, some energy is dissipated as heat and is no longer available to do work. Although the total amount of energy in the universe is not decreasing with time, the energy available to do work does decrease because useful forms of energy are continuously degraded to heat, a less useful form of energy.

In fact, heat is actually the energy of random molecular motion. Heat can be made to do work when there is a temperature gradient, that is, a difference in potential. The term *entropy* refers to the energy that has become so randomized and uniform throughout a system that it is no longer available to do work.

It is a measure of the disorder of a system. According to the second law of thermodynamics, entropy in the universe is continuously increasing. Eventually all energy will be random and uniform in distribution. With only this useless form of energy, no work could be performed. The universe will have run down. We are in no immediate danger, however, because this depressing state is not scheduled to occur for several billion years.

Because of the second law of thermodynamics, no process requiring energy is ever 100% efficient. Much of the energy is dispersed as heat and is transferred to the environment, so that there is an increase in

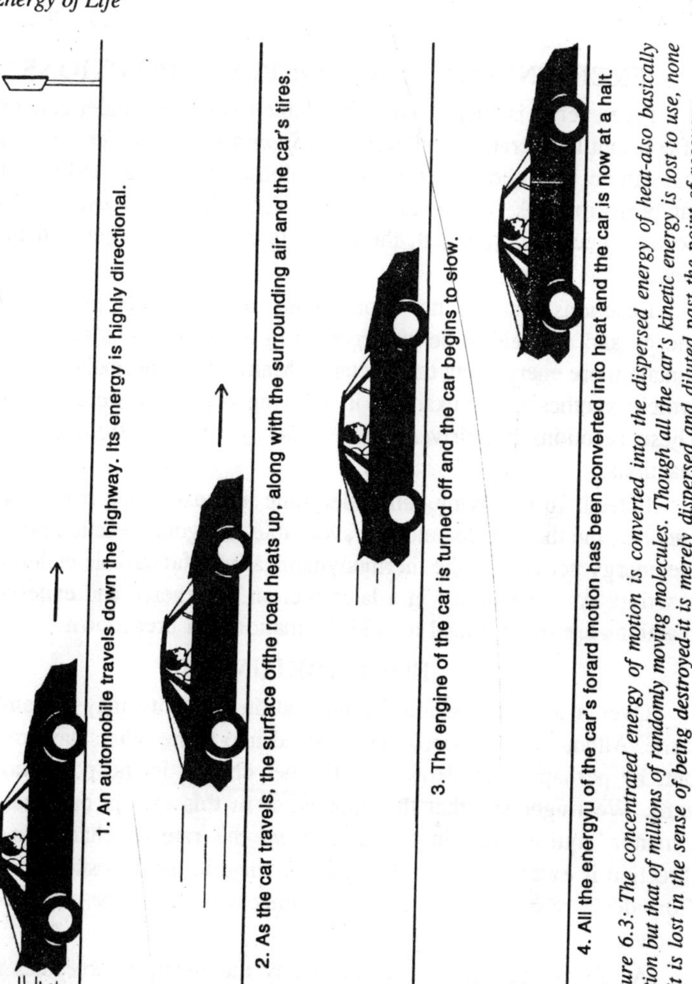

Figure 6.3: The concentrated energy of motion is converted into the dispersed energy of heat-also basically motion but that of millions of randomly moving molecules. Though all the car's kinetic energy is lost to use, none of it is lost in the sense of being destroyed-it is merely dispersed and diluted past the point of recovery.

entropy. Cellular energy utilisation is about 55% efficient, with the other 45% of the energy being lost as heat. Such biological processes are actually quite efficient compared with most machines made by human beings. For example, a gasoline engine is only about 17% efficient.

Because they are highly organised, living organisms are very unstable. In fact, life is a constant struggle against the second law of thermodynamics. Survival of individual organisms, as well as ecosystems, depends upon continuous energy input. Thus, producers must carry on photosynthesis, and consumers and decomposers must eat.

FREE ENERGY AND COUPLED REACTIONS

Free energy is energy that is available to do work under conditions of constant temperature and pressure. *Spontaneous reactions*, reactions that can occur without outside input, release free energy and therefore can perform work. These reactions are also called *exergonic reactions*. Because energy is released, the products contain less energy than the reactants.

Reactions that are not spontaneous require a net input of free energy and are said to be *endergonic*. In these reactions the products contain more energy than the reactants. Many of the metabolic reactions-protein synthesis, for example in a living organism are endergonic. These reactions are driven by the energy released from exergonic reactions.

Indeed, in the living cell endergonic and exergonic reactions are *coupled*; the thermodynamically favourable exergonic reaction provides the energy needed for the thermodynamically unfavourable endergonic reaction. As we will see in a later section, exergonic and endergonic reactions are often linked to ATP formation and breakdown.

EQUILIBRIUM

Over a ten-year period the population of a city may remain the same. Although some people have moved into town, others have moved out, or perhaps died. However, the net change in the population is zero. We might say that the population in this city is in a state of dynamic equilibrium. In an *equilibrium* the rate of change in one direction is exactly equal to the rate of change in the opposite direction. Consider a reaction in terms of the numbers of each type of molecule involved.

At the beginning of a reaction only the reactant molecules may be present. These molecules move about and collide with one another with sufficient energy to react. As more and more product molecules are produced, fewer and fewer reactant molecules are left. As the number of product molecules increases, they collide more frequently, and some have sufficient energy to initiate the reverse reaction.

The reaction proceeds in both directions simultaneously and reaches an equilibrium in which the rate of the reverse reaction is about the same as the rate of the forward reaction. Reversibility of chemical reactions permits cells to control their release of free energy according to their needs. It also allows cells to resynthesize many large compounds for use in metabolism.

When the free energy difference between the products and reactants of a chemical reaction is zero, the reaction is at equilibrium. Any change, such as a change in temperature or pressure, that affects the reacting system may cause the equilibrium to shift. The reaction may then proceed in a specific direction until once again the free-energy difference is zero and a new equilibrium has been established.

When there is little free-energy difference on the two sides of a chemical equation, the direction of a reaction will be governed mainly by the concentrations of the reactants and products. The reaction tends to proceed in the direction that will minimize the difference in concentration between the substances on the two sides of the equation. If one substance, say, the product, is continuously removed as it is formed, the reaction will indeed proceed to completion.

ENZYMES: CHEMICAL REGULATORS

The thousands of chemical reactions that take place within the living cell must be precisely controlled and coordinated. These reactions interlock somewhat as a multitude of assembly processes interlock on a modernday production line. In the manufacturing of an automobile, for example, it might not be possible to complete a car without some seemingly minor part, such as a ball bearing or special bolt.

Very elaborate planning and scheduling are necessary to ensure that all parts are delivered to the plant by the time they are needed and that each worker performs a small part of the overall job at precisely the right time. It is this concept of control that we wish to emphasize. Although energy can be released from glucose and other fuel molecules by burning, cells could not survive such high temperatures nor effectively utilize energy that is released in short sudden bursts. Cells require a slow, steady release of energy that they must be able to regulate to meet metabolic energy requirements.

Accordingly, fuel molecules are slowly oxidized and energy is extracted in small amounts during cellular respiration, which includes sequences of 30 or more reactions. In fact, most cellular metabolism proceeds by a series of steps, so a given molecule may go through as many as 20 or 30 chemical transformations before it reaches some final state. And then the apparently completed molecule may be pre-empted by yet another chemical pathway so as to be totally transformed or consumed in the course of metabolism. The changing needs of the cell require a system of flexible chemical control. The key elements of this control system are the remarkable enzymes.

What Enzymes Do

An *enzyme* is an organic catalyst, a compound that greatly increases the speed of a chemical reaction without being consumed itself. All enzymes are proteins (though, of course, not all proteins are enzymes). Like all catalysts, an enzyme affects the rate of a reaction by lowering the energy needed to activate the reaction.

Even a strongly exergonic reaction that releases more than enough energy as it proceeds is prevented from beginning by an energy barrier. This is because before new chemical bonds can be formed, existing ones must be broken. The energy required to overcome this barrier and get the reaction going is called *activation energy*.

An enzyme greatly reduces the activation energy necessary to initiate a chemical reaction. An enzyme can only promote a chemical reaction that could be made to proceed without it. There is nothing in the action of a catalyst of any kind that could change the operation of the second law of thermodynamics, so enzymes do not influence the direction of a chemical reaction or the final concentrations of the molecules involved.

They simply *speed* up reactions. Consider the enzyme carbonic anhydrase as an example of what enzymes do. Like most enzymes it is identifiable as an enzyme by its distinctive **-ase** ending. Carbonic anhydrase promotes the interconversion of carbonic acid with carbon dioxide and water:

$$\underset{\text{Carbon dioxide}}{CO_2} + \underset{\text{Water}}{H_2O} \xrightleftharpoons{\text{carbonic anhydrase}} \underset{\text{Carbonic acid}}{H_2CO_3}$$

This reaction takes place in many cells and tissues, especially in the kidneys and red blood cells. Carbonic acid ionizes, forming bicarbonate ions, the form in which most of the carbon dioxide in the blood is transported. The reaction shown above could proceed without the enzyme but would be uselessly slow. When carbonic anhydrase is present, the reaction proceeds about 10 million times faster. Amazingly, a single molecule of this enzyme can promote the conversion of an estimated 600,000 molecules of carbon dioxide into carbonic acid each second.

How Enzymes Work

The chemicals upon which an enzyme operates are referred to as its *substrates*. Enzymes are very *specific*. An enzyme will catalyze only a few closely related chemical reactions, or in many cases only one particular reaction. Enzymes form temporary chemical compounds with their substrates. These complexes then break up, releasing the product and regenerating the original enzyme molecule for reuse.

Enzyme + Substrate 1 + Substrate 2

→ Enzyme-substrate complex

Enzyme-substrate complex → Enzyme + Product(s)

Note that the enzyme itself is not permanently altered or consumed by the reaction.

Why does the enzyme-substrate complex break up into chemical products different from those that participated in its formation? As shown in Figure elsewhere in this chapter, each enzyme has one or more regions called *active sites*, which in the case of a few enzymes have been shown to be actual indentations in the enzyme molecule.

These active sites are located close to one another on the enzyme's surface, so during the course of a reaction, substrate molecules occupying these sites are temporarily brought together and react with one another. It is thought that when the enzyme and substrate bind together, the shape of the enzyme molecule changes slightly.

This produces strain in critical bonds in the substrate molecules so that these bonds break. The new chemical compound thus formed has little affinity for the enzyme and moves away from it. An enzyme can be thought of as a molecular lock into which only specifically shaped molecular keys-the substrates-can fit. Unlike a lock and key, however, the enzyme and its substrate seem not to be exactly complementary shapes.

A recent model of enzyme action, known as the *induced-fit model*, is based on data indicating that the active sites of an enzyme are not rigid. When the substrate binds to the enzyme, it may induce a change in shape in the enzyme molecule, resulting in an optimum fit for the substrate-enzyme interaction. The change in shape of the enzyme molecule can put strain on the substrate. This stress may help bonds to break, thus promoting the reaction.

Some enzymes-for example, pepsin, secreted by the stomachconsist only of protein. Other enzymes have two components, a protein referred to as the apoenzyme and an additional chemical component called a *cofactor*. The cofactor of some enzymes is a metal ion that is often a functional part of the active site. Many of the trace elements required by organisms function as cofactors. Together, the cofactor and apoenzyme work very efficiently. Hydrogen peroxide can be split by inorganic catalysts, such as free iron atoms. However, it would take

300 years for an iron atom to split the same number of molecules of hydrogen peroxide split in one second by a molecule of the enzyme catalase, which contains a single iron atom. An organic, nonpolypeptide compound that serves as a cofactor is called a *coenzyme*. Many coenzymes are synthesized from vitamins, particularly from the B vitamins. The coenzyme serves as an adapter, permitting the enzyme to accept a substrate for which it would, by itself, have little affinity.

Regulating Enzymatic Action

Enzymes regulate the chemistry of the cell, but what controls the enzymes? One mechanism of enzyme control simply depends upon the amount of enzyme produced. The synthesis of each type of enzyme is directed by a specific gene. The gene, in turn, may be switched on by a signal from a hormone or by some other type of cellular product. When the gene is switched on, the enzyme is synthesized.

Then the amount of enzyme present influences the rate of the reaction. Up to a maximum value, the rate of an enzyme-dependent reaction increases as the concentration of the enzyme increases. Another important method of enzymatic control depends upon the activation of enzyme molecules that are present in an inactive form in the cytoplasm.

In the inactive form the active sites of the enzyme are inappropriately shaped, so the substrates do not fit. Among the factors that influence the shape (conformation) of the enzyme are acidity and alkalinity and the concentration of certain salts.

Some enzymes possess an activator site (also called an allosteric site). When a molecular activator, such as the substance cyclic AMP, occupies the activator site, the shape of the enzyme molecule changes, making the active sites better suited for binding with the substrate.

Enzyme Optima

Enzymes generally work best under certain narrowly defined conditions referred to as optima. These include appropriate temperature, pH, and salt concentration. For example, pepsin, the protein-digesting enzyme of the stomach, works best at the strongly acid pH of 2. In contrast, the starch-digesting enzyme amylase in saliva and pancreatic juice has a pH optimum of 8.5 (slightly alkaline). Strong acids or bases irreversibly inactivate most enzymes by permanently changing their molecular conformation.

Enzymatic reactions occur very slowly or not at all at low temperatures, but activity resumes when the temperature is raised to normal. The rates of most enzyme-regulated reactions increase with increasing temperature, within limits. Temperatures greater than 50° or 60° C rapidly inactivate most enzymes by altering their secondary and tertiary level structures. The enzyme is said to be denatured. (An everyday example of this is cooking an egg white; it changes in consistency as the protein is denatured.)

Enzyme Inhibition

Enzymes may be rendered less effective or even nonfunctional by certain chemical agents known as *inhibitors*. Enzyme inhibition may be reversible or irreversible. Reversible inhibitors can be competitive or noncompetitive. In *competitive inhibition* the inhibitor competes with the normal substrate for the active site of the enzyme.

A competitive inhibitor usually is chemically similar to the normal substrate and so fits the active site and binds with the enzyme. However, it is not similar enough to the normal substrate to take its place effectively. The enzyme cannot act upon it to form reaction products. A competitive inhibitor occupies the active site only temporarily and does not damage the enzyme irreversibly.

In *noncompetitive inhibition* the inhibitor binds with the enzyme at a site other than the active site. Such an inhibitor renders the enzyme inactive by altering its shape. Many important noncompetitive inhibitors are metabolic substances that help regulate enzyme activity by combining reversibly with the enzyme. Many poisons are *irreversible inhibitors* that permanently inactivate or even destroy the enzyme.

Nerve gases are irreversible inhibitors that poison the enzyme cholinesterase, essential in the normal function of nerves and muscles. A number of insecticides and drugs are irreversible inhibitors. The antibiotic penicillin and its chemical relatives inhibit a bacterial enzyme necessary for bacterial cell wall construction.

Unable to produce new cell walls, susceptible bacteria cannot multiply effectively. Since human body cells do not have cell walls (and so do not employ the susceptible enzyme), penicillin is harmless to humans, except for the occasional allergic patient.

PACKAGING ENERGY—ATP

In all living cells energy is temporarily packaged within a remarkable chemical compound called *adenosine triphosphate*, or more simply, ATP. This compound stores large amounts of energy for very short periods of time. We may think of ATP as the energy currency of the cell. When you work, you earn money, and so you might say that your energy is symbolically stored in the money you earn. In the same way, the energy of the cell may be stored in ATP. When you have earned extra money, you might deposit some in the bank. Similarly, the cell deposits energy as lipid in fat cells or as glycogen in the liver and muscle. Moreover, just as you dare not make less money than you spend, so, too, the cell must avoid energy bankruptcy, which would mean its death. Finally, just as you (alas) do not keep what you make very long, so too the cell is forever spending its ATP.

ATP, a nucleotide, consists of three main parts:

(1) a nitrogen-containing base, adenine, which also occurs in DNA,

(2) ribose, a five-carbon sugar, and

(3) three inorganic phosphate groups that are identifiable as phosphorus atoms surrounded by oxygen atoms.

Inorganic phosphate is usually designated P_i. Notice that the phosphate groups are attached to the end of the molecule in a series, rather like three passenger cars behind a locomotive. The couplings, that is, the chemical bonds attaching the last two phosphates, also resemble those of a train in that they are readily attached and detached.

When the third phosphate is removed, the remaining molecule is called *adenosine diphosphate*, or simply ADP. When two phosphate groups are removed, the molecule that remains is *adenosine monophosphate*, AMP. Conversely, when a phosphate is attached to AMP, it becomes ADP, and when a phosphate is added to ADP, ATP is produced. Addition of a phosphate group to a molecule is referred to as *phosphorylation*. Note that these reactions are readily reversible.

$$AMP + P_i + Energy \rightleftarrows ADP$$

$$ADP + P_i + Energy \rightleftarrows ATP$$

As these equations indicate, energy is required to add a phosphate to either the AMP or the ADP molecule. Converselyu since energy can neither be created nor destroyed, the energy is released or transferred to another molecule when the phosphate is detached. Thus ATP is an important link between exergonic (energy-releasing) and endergonic (energy-requiring) reactions.

The high energy bonds in ATP are often indicated by wavy lines attaching the last two phosphates to the rest of the molecule. These bonds contain more than twice the energy of an average chemical bond because an unusually large amount of energy is required to join the second and third phosphates.

Energy released from exergonic reactions is packaged in ATP molecules for use in endergonic reactions. This energy may be used to produce fats or glycogen, molecules stockpiled for long-term energy storage. The cell contains a pool of ADP, ATP, and phosphate in a state of equilibrium. Large quantities of ATP cannot be stockpiled in the cell. In fact, recent studies suggest that a bacterial cell has no more than a one-second supply of ATP.

Thus ATP molecules are used almost as quickly as they are produced. Every second, thousands of ATPs are made from ADP and phosphate and an equal number are broken down, yielding their energy to whichever life processes may require them. In this way ADP and phosphate are ceaselessly recycled, shuttling back and forth in a blur of energy between cellular respiration and consumption.

Most ATP is manufactured during the electron transport phase of cellular respiration in the mitochondria. However, ATP can also be produced at what is called the substrate level. Imagine that a particular substrate is chemically combined with phosphate to form substrate-phosphate, which with the help of an enzyme, substrate-phosphatase, can donate its phosphate to ADP. In accepting the phosphate, ADP becomes ATP.

$$\text{Substrate-phosphate} + \text{ADP} \xrightarrow[\text{phosphatase}]{\text{Substrate}} \text{Substrate} + \text{ATP}$$

Why does the phosphate leave the original substrate compound and join ADP? Chemical reactions, according to the second law of thermodynamics, tend to proceed in the direction of higher to lower potential energy. Thus, because there is more potential energy in the substratephosphate than in ATP, the reaction proceeds toward the production of ATP. Only a few compounds in the cell meet these requirements, however, so not much ATP is synthesized in this way. Still, muscle cells make considerable use of this method of ATP production.

HYDROGEN AND ELECTRON TRANSFER

Energy processing by cells involves the transfer of chemical energy through the flow of electrons and protons. Generally, there is a sequence of redox (oxidation-reduction) reactions that take place as hydrogen or its electrons are transferred from one compound to another. Oxidation is a chemical reaction in which a substance loses electrons. In living cells oxidation almost always involves the removal of a hydrogen atom (which contains an electron) from a compound. Reduction is a chemical reaction in which a substance gains electrons (or hydrogens).

Electrons released during an oxidation reaction cannot exist in the free state in living cells. That is why every oxidation must be accompanied

Figure 6.4: The total energy released by a falling object is the same whether it is released all at once or in a series of steps. Similarly, in cellular metabolism the energy of an electron liberated from a foodstuff is the same whethere it is released all at once, or, as is actullly the case, gradually as it passes to successive electron acceptors.

by a reduction in which the electrons are accepted by another atom or molecule. The oxidised molecule gives up energy, while the reduced molecule receives energy. When hydrogen atoms are removed from an organic compound, they take with them some of the energy that had been stored in their chemical bonds.

The hydrogen along with its energy is transferred to a hydrogen acceptor molecule, which is generally a coenzyme. One of the most frequently encountered hydrogen acceptor coenzymes is *nicotinamide adenine dinucleotide*, more conveniently referred to as NAD. NAD is a coenzyme that can temporarily package large amounts of free energy. Here is a generalized equation showing the transfer of hydrogen from a compound we will call X to NAD:

$$XH_2 + NAD^+ \rightarrow \underset{\text{Oxidised}}{X} + \underset{\text{Reduced}}{NAD - H} + H^+$$

Note that the NAD is reduced when it combines with hydrogen. Some of the energy stored in the bonds holding the hydrogens to molecule X has been transferred by this reaction to the NADH. This energy can now be used for some metabolic process or it can be transferred through a complex series of reactions to ATP. Transfer of energy in both photosynthesis and cellular respiration involves sequences of redox reactions. The acceptor compounds that make up these sequences keep energy flowing.

Other important hydrogen or electron acceptor compounds are *flavin adenine dinucleotide (FAD)* and the cytochromes. FAD is a nucleotide that accepts hydrogens and their electrons. The cytochromes are proteins that contain iron. The iron component accepts electrons from hydrogen and then transfers them to some other compound. Cytochromes, like NAD and FAD, are important in cellular respiration. Such electron acceptors provide a mechanism by which the cell can slowly and efficiently capture energy from fuel molecules.

7

Capturing Energy by Cells

Our technological society will someday run out of fossil fuels. What energy sources will be used then to power our industries, homes, and vehicles? One seriously discussed alternative source of energy is hydrogen. However, there are practical difficulties in storing and transporting this highly reactive gas, so hydrogen may have to be combined with carbon to form artificial hydrocarbon fuels similar to those already in use, such as gasoline or propane. This would involve some sacrifice in energy content, but since these fuels have a high hydrogen content, they would still serve well, as indeed they do now.

Even if we had no carbon in elemental form with which to combine the hydrogen, hydrogen-carrying fuels could still be produced, at some further energy sacrifice, from such common carbon sources as carbon dioxide. Modification of the process would permit the production of still other energy-rich substances, such as alcohol or glycerine, which are important in industry. Although making practical use of hydrogen fuel has not yet become a reality in human society, the biological world has long made use of this energy source.

The vast chemical industry of the biosphere has for millennia captured solar energy by photolysis (breaking down with light energy) of water, released the resulting oxygen, and incorporated the hydrogen into energy-rich carrier compounds. These compounds form the matter of biological commerce because they also can be incorporated into the body structure of organisms as structural components. The energy of these hydrogen-containing compounds can be released once again by allowing them to react once more with the formerly released oxygen. The details of these metabolic processes form the subject of this chapter.

Plants and algae are producers that obtain their energy from the sun. They are uniquely capable of converting solar energy into stored chemical energy-the process of *photosynthesis*. Each year these remarkable organisms produce more than 200 billion tons of food.

The chemical energy stored in this food fuels the chemical reactions that sustain life. All consumers obtain their energy by consuming plants and algae or by eating other consumers that have fed on plants and algae. Almost all life, therefore, is ultimately dependent upon photosynthesis. Chemical energy is a stable energy form; it may be stored for long periods of time. A striking example is the energy in coal and oil, which was stored from photosynthesis millions of years ago.

LIGHT AND LIFE

Since life on our planet depends upon light, it seems appropriate to discuss the nature of light and how it permits photosynthesis to occur. Light is a very small portion of a vast, continuous spectrum of radiation, the electromagnetic spectrum. All radiations in this spectrum behave as though they travel in waves. At the end of the spectrum are gamma rays with very short wavelengths (measured in nanometers).

A *wavelength* is the distance from one wave peak to the next. At the other end of the spectrum are low-frequency radio waves, with wavelengths so long that they are measured in kilometers. Within the spectrum of visible light, violet has the shortest wavelength and red the longest. Ultraviolet light has a still shorter range of wavelengths and infrared a still longer one. Light behaves not only as a wave but also as a particle. It is composed of particles of energy called *photons*.

The energy of a photon is different for light of different wavelengths. The shorter the wavelength, the more energy light has, and the longer the wavelength, the lower the energy. In other words, the energy of the photon is inversely proportional to the wavelength.

Why does photosynthesis depend upon visible light rather than upon some other wavelength of radiation? We can only speculate, but one reason may be that most of the radiation reaching our planet from the sun is within this portion of the electromagnetic spectrum. Another consideration is that only radiation within the visible light portion of the spectrum excites certain types of biological molecules, moving electrons into higher energy levels.

Photons interact with atoms in a variety of ways, but all the ways depend on the electron structure of the atom. As you should recall, an

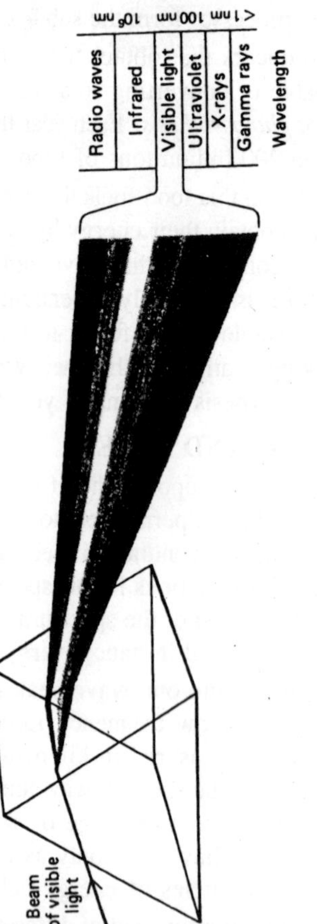

Figure 7.1: Visible light is only a portion of the electromagnetic spectrum. The green pigment chlorophyll absorbs light mainly in the blue, violet, and red regions of the spectrum. During photosynthesis light energy is used to synthesize organic compounds.

atom consists of an atomic nucleus surrounded by one or more energy levels containing electrons. In the hydrogen atom there is only a single electron that occupies the first energy level. The lowest energy state an atom possesses is called the *ground state*, but energy can be added to an electron so that it will attain a higher energy level. When an electron is raised to a higher energy level than its ground level, the atom is said to be excited.

When an electron is raised to a higher energy state it may soon return to its ground level. In this case energy is usually dissipated as heat or as light of a longer wavelength. This is referred to as fluorescence. Or the excited electron may be lost, leaving the atom with a net positive charge. In this instance the electron may be accepted by a reducing agent. This is what occurs in photosynthesis.

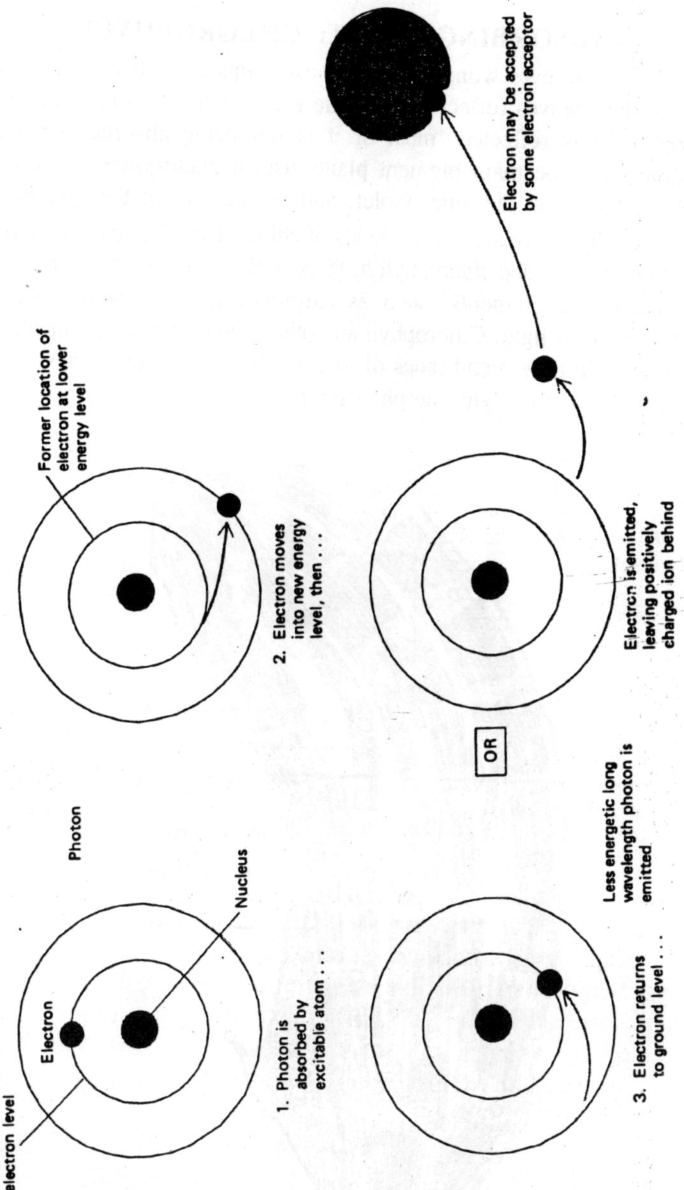

Figure 7.2: A quantum of light energy strikes an atomor the molecule of which the atom is a part. The energy of the photon may push the electron to an orbit farther from the nucleus.

ABSORBING LIGHT: CHLOROPHYLL

Have you ever wondered why most plants are green? The reason is that their leaves reflect most of the green light that strikes them. If green light is reflected, most of it is not being absorbed or used. Chlorophyll, the main pigment plants use in photosynthesis, absorbs light primarily in the blue, violet, and red regions of the spectrum.

Actually, there are several kinds of chlorophyll. The most important are chlorophyll *a* and chlorophyll *b*. Plants (and algae) also have accessory photosynthetic pigments, such as carotenoids, that absorb different wavelengths of light. Chlorophyll and other photosynthetic pigments are located within the membranes of *thylakoids*, tiny flattened sacs within cells capable of carrying on photosynthesis.

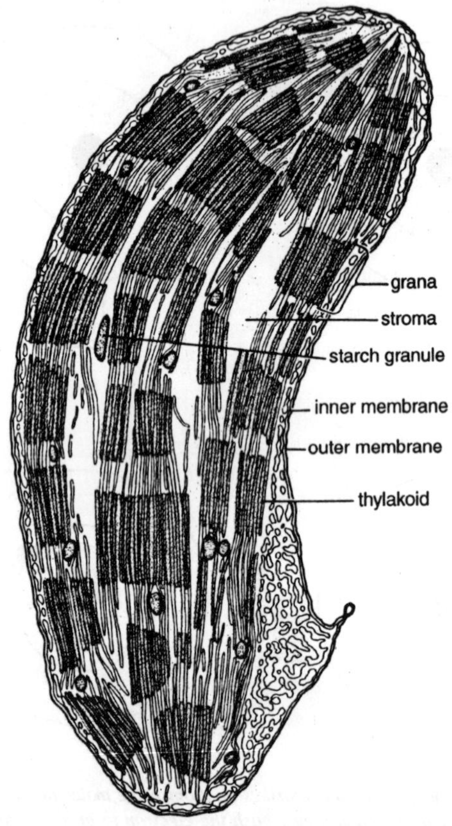

Figure 7.3: Electron micrograph and diagram of a chloroplast from a leaf cell of a timothy plant, Phleum pratense. *Note grana, which consist of stacks of thylakoids.*

In prokaryotes that photosynthesize, thylakoids often occur as extensions of the cell membrane and may be arranged around the periphery of the cell. In photosynthetic eukaryotes, thylakoids are found within chloroplasts. All the enzymes required for the light-dependent reactions of photosynthesis are associated with the thylakoid membranes.

Let us examine a chloroplast and its component thylakoids. Surrounded by two membranes, the chloroplast has an interior packed with stacks of thylakoids. These stacks are referred to as *grana*. Each granum looks something like a stack of coins, and each "coin" is a thylakoid. Some thylakoid membranes extend from one granum to another.

The fluid-filled region of the chloroplast outside the thylakoids is called the *stroma*. Most of the enzymes required for the light-independent reactions of photosynthesis are found in the stroma.

THE REACTIONS OF PHOTOSYNTHESIS

The principal raw materials for photosynthesis are water and carbon dioxide. Using the energy that chlorophyll molecules trap from sunlight, water is split, its oxygen is liberated, and its hydrogen is combined with carbon dioxide to produce carbohydrate molecules. The reactions of photosynthesis may be summarised as follows:

$$\underset{\text{Carbon dioxide}}{6CO_2} + \underset{\text{Water}}{12H_2O} \xrightarrow[\text{chlorophyll}]{\text{Light energy}} \underset{\text{Glucose}}{C_6H_{12}O_6} + \underset{\text{Oxygen}}{6O_2} + \underset{\text{Water}}{6H_2O}$$

This equation describes what happens during photosynthesis but not how it happens. The "how" is much more complex and involves many chemical reactions. Although a detailed examination of photosynthesis is beyond the scope of a beginning biology course, we will describe the principal reactions. For convenience we can divide the reactions of photosynthesis into the light-dependent and the lightindependent reactions.

The Light-Dependent Reactions

The light-dependent reactions can take place only in the presence of light. During this phase of photosynthesis several important events occur.

1. Chlorophyll absorbs light energy, which is immediately converted to electrical energy. The electrical energy is represented in the flow of electrons from the chlorophyll molecule. We say that the chlorophyll becomes temporarily energised.

Table 7.1: Summary of Principal Reactions of Photosynthesis.

Reaction series	Summary of process	Needed materials	End products
Light-dependent reactions (take place in thylakoid membranes)	Energy from sunlight used to split water, manufacture ATP, and reduce NADP		
(1) photochemical reactions	Chlorophyll energised; reaction center gives up energised electron to electron acceptor	Light energy; pigments, such as chlorophyll	Electrons
(2) electron transport	Electrons are transported along chain of electron acceptors in thylakoid membranes; electrons reduce NADP; splitting of water provides some of H^+ that accumulates inside thylakoids	Electrons, NADP, H_2O, electron acceptors	NADPH + H^+, O_2, H^+
(3) chemiosmosis	HT are permitted to move across the thylakoid membrane down proton gradient; they cross membrane through	Proton gradient, ADP + P_i	ATP

Table 7.1 (contd.)

Table 7.1 (contd.)

Reaction series	Summary of process	Needed materials	End products
	special channels formed by protein complex CF_0-CF_1; energy released is used to produce ATP		
Light-independent reactions (take place in stroma)	Carbon fixation; carbon dioxide is combined with organic compound	Ribulose bisphosphate, CO_2, ATP, NADPH + H^+	Carbohydrates, ADP + P_i, $NADP^+$

2. Some of the energy of the energised chlorophyll is used to make ATP. During this process electrical energy is transformed to chemical energy.
3. Some of the photon energy trapped by the chlorophyll is used to split water, a process known as *photolysis*. Oxygen from the water is released. Some of the oxygen is used by the plant for cellular respiration, but most of it is released into the atmosphere.
4. Hydrogen from the water combines with the hydrogen carrier molecule $NADP^+$, forming NADPH (reduced NAD). Here again, electrical energy is converted to chemical energy.

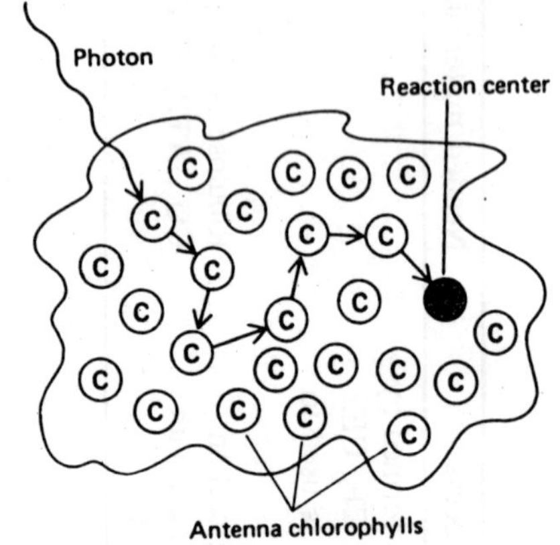

Figure 7.4: A photosynthetic unit. The many chlorophyll molecules (c) in the unit are excited by photons and transfer their excitation energy to the specially positioned chlorophyll molecule, the reaction center.

We may summarize the light-dependent reactions as follows:

$$12H_2O + 12NADP + 18ADP + 18P_i \xrightarrow[\text{chlorophyll}]{\text{Light energy}} 6O_2 + 12NADPH + 12H^+ + 18ATP$$

Thus, in the light-dependent reactions the energy from sunlight is used to make ATP and to reduce NADP. Some of the captured energy is temporarily stored within these two compounds.

The Photosystems

According to the currently accepted model, chlorophyll molecules and associated electron acceptors are organised into units called *photosystems*. There are two photosystems, each containing up to 400 molecules of chlorophyll. Photosystem I contains a reactive pigment (perhaps a special form of chlorophyll a) known as P700 because one of the peaks of its light absorption spectrum is at 700 nanometers. Photosystem II utilizes a pigment, P680, whose absorption maximum is at a wavelength of 680 nanometers.

All chlorophyll molecules of a photosystem apparently serve as antennae to gather solar energy. When they absorb light energy, it is passed from one chlorophyll molecule to another until it reaches the special P700 or P680 pigment molecule, referred to as the *reaction center*. Only this molecule is able to give up its energised electron to an electron acceptor compound.

In photosystem I the energised electron is first transferred to a compound known as bound ferredoxin, then to a second electron acceptor, and finally to the electron acceptor $NADP^+$. When $NADP^+$ is in an oxidised acceptance condition, it is positively charged. When it accepts electrons, the electrons unite with protons to form hydrogen, so the reduced form of $NADP^+$ is NADPH. The reduction of $NADP^+$ requires two electrons. Thus two photons of light must be absorbed by photosystem I to form one NADPH. (Ferredoxin, however, can only transfer one electron at a time.)

When an electron is transferred to ferredoxin, photosystem I becomes positively charged. It could never emit another electron, whether or not it became excited, unless one somehow were first restored to it. That needed electron is donated by photosystem II. Like photosystem I, photosystem II is activated by a photon and gives up an electron to a chain of electron acceptors. The missing electrons are replaced by electrons from water. When P680 absorbs light energy, it becomes positively charged and exerts a strong pull on the electrons in water molecules.

Water is split (photolysis) into its components-protons (H^+), electrons, and oxygen. The protons from water are transferred to NADP, which explains how the NADP is converted to NADPH + H^+. The oxygen split from the water is released into the atmosphere.

The electron emitted from photosystem II passes from one acceptor to another through a chain of easily oxidised and reduced compounds. (Among these are a coenzyme related to vitamin K; iron-containing pigments known as cytochromes; and a copper-containing protein.) As electrons are transferred along this chain of electron acceptors, they become less and less energised. Some of the energy released is used to establish a proton gradient, which leads to the synthesis of ATP (see next section). An electron emitted from photosystem II is eventually donated to photosystem I. The process just described is known as *noncyclic photophosphorylation*. It is not a cycle because there is a one-way flow of electrons to

NADP, and it is called photophosphorylation because electrons obtain energy from photons and then contribute that energy to the phosphorylation of ADP (producing ATP). For every two electrons that enter the pathway of noncyclic photophosphorylation, there is an energy yield of two ATP molecules and one NADPH.

Cyclic Photophosphorylation

Only photosystem I is involved in cyclic photophosphorylation. In this pathway the electrons originate from photosystem I and eventually are returned to the same photosystem. For every two electrons that enter the pathway, one ATP molecule is produced. NADPH is not produced, which means that there is a far lower energy yield and that carbon dioxide is not reduced to produce carbohydrates. Photolysis does not occur in cyclic photophosphorylation, so oxygen is not produced in this process.

Cyclic photophosphorylation occurs in plant cells when there is too little NADP to accept electrons from ferredoxin. Biologists think that this process was used by ancient bacteria to produce ATP from light energy. A reaction pathway analogous to cyclic photophosphorylation in plants is present in modern photosynthetic bacteria.

How Proton Gradients Form ATP: Chemiosmosis

The photosystems and electron acceptors are embedded in the thylakoid membrane. According to the *chemiosmotic theory*, energy released from electrons traveling through the chain of acceptors is used to pump protons across the thylakoid membrane into the thylakoid. These protons accumulate within the thylakoids. Because protons and hydrogen ions (H^+) are the same, this accumulation of protons causes the pH of the thylakoid interior to fall. In fact, the pH approaches 4 in bright light. This produces a pH difference of about 3.5 units across the thylakoid membrane-more than a thousand-fold difference in hydrogen ion concentration.

In accordance with the general principles of diffusion, the highly concentrated hydrogen ions inside the thylakoid tend to diffuse out. However, they are prevented from doing so because the thylakoid membrane is impermeable to them except at certain points bridged by a remarkable protein, the **CF_0-CF_1 complex,** also called *ATP synthetase*. This complex extends across the thylakoid membrane, projecting from

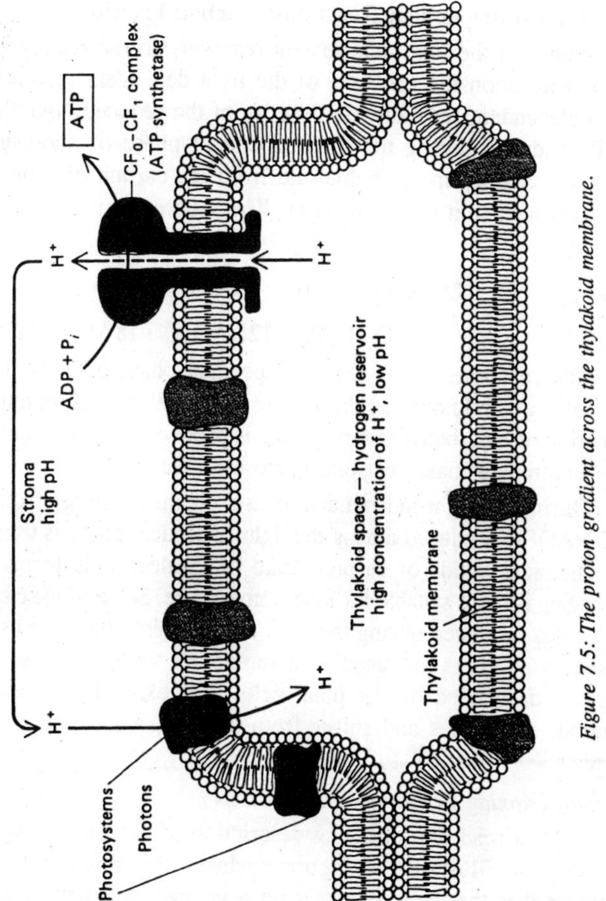

Figure 7.5: The proton gradient across the thylakoid membrane.

the membrane surface both inside and outside. These proteins actually form channels through which protons can leak out of the thylakoids.

Note that the concentrated proton solution inside the thylakoids represents a low entropy state. If the protons can move out of the thylakoids into the surrounding space, the resulting random and more or less even distribution of protons within the chloroplast would be a high entropy state. The second law of thermodynamics identifies this situation as one in which there is a potential for useful work resulting from the change in entropy. What happens is that as the protons pass through the CF_0-CF_1 complex, energy is released, and in some way the complex causes ATP to be synthesized. Just how this is done is now the subject of active investigation.

The Light-Independent Reactions: Carbon Fixation

Although the light-independent reactions do not *require* light, they do depend upon the products of the light-dependent reactions. In the lightindependent reactions, the energy of the reduced NADPH and the ATP produced during the light-dependent phase of photosynthesis is used to manufacture carbohydrate from the carbon dioxide of the air and hydrogen split off from water. The process may be summarised as follows:

$$12NADPH + 12H^+ + 18ATP + 6CO_2 \rightarrow$$

$$C_6H_{12}O_6 + 12NADP^+ + 18ADP + 18P_i + 6H_2O$$

The reactions of the light-independent phase actually proceed by way of a cycle known as the *Calvin cycle.* With each complete cycle a portion of a carbohydrate molecule is produced. Six turns of the cycle are required to make just one glucose molecule.

During the light-independent phase, chemical energy from the ATP and NADPH produced during the light-dependent phase is transferred to the chemical bonds of carbohydrate molecules. This form of energy packaging is more suitable for long-term storage. Some of the carbohydrate molecules produced during the light-independent phase are used as fuel molecules. Others are used to manufacture various types of organic compounds needed by the plant cells. For example, by adding such minerals as nitrates and sulfur from the soil, plant cells can produce proteins.

Carbon Dioxide Fixation: Some Details

As you read the following description of the Calvin cycle, follow the reactions illustrated in Figure elsehwhere in this chapter. We will consider that the cycle begins with a pentose sugar phosphate called *ribulose phosphate*. In the first chemical transformation, ATP from the light-dependent reactions is expended to add a second phosphate to the five-carbon skeletons of three ribulose phosphates. This chemical reaction converts them to *ribulose bisphosphate* (RuBP). A key enzyme then combines three molecules of carbon dioxide with three molecules of RuBP. This is called *CO_2 fixation*.

Instantly each molecule splits into two three-carbon molecules called phosphoglycerate (PGA). There are now six three-carbon molecules. With the energy from more ATP from the light-dependent phase the PGA molecules are converted to molecules of diphosphoglycerate (DPGA). Next, DPGA is converted to *glyceraldehyde-3-phosphate*, known simply as PGAL. In this conversion hydrogen is donated by reduced NADP

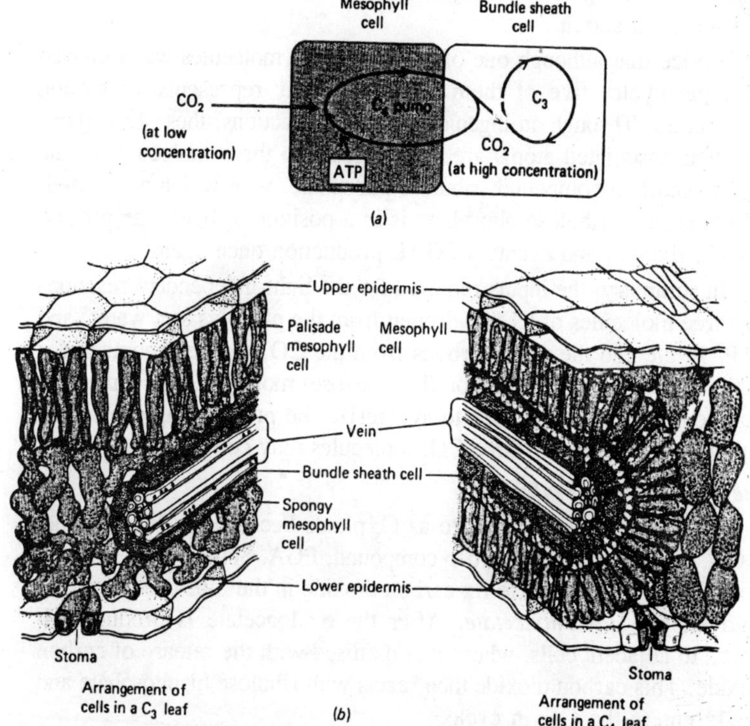

Figure 7.6: (a) Overview of the C_4 series of reactions in photosynthesis. If, as usual, CO_2 is present in low concentration, the C_4 system readily absorbs it and in effect concentrates it for use by the C_3 system to which it is pumped. Since the C_4 system consumes some energy, ultimately made available only by photosynthesis, this system is important to the plant only at high light intensities when the stomata can be kept open despite those high light intensities. Thus, to be effective, the C_4 system requires an abundance of both water and light. Yet under these conditions, it can fix more carbon than the C_3 system can fix by itself. (b) Comparison of the leaf structures of a C_3 and a C_4 plant.

formed during the light-dependent reactions. Now the plant cell is in a position to harvest the three CO_2 molecules it has put into the process, not in their original form but as carbohydrate.

One of the six PGAL molecules leaves the system to be used in carbohydrate synthesis and perhaps ultimately in the synthesis of other organic molecules, such as proteins or fats. Each of these three-carbon molecules of PGAL is essentially half a hexose (six-carbon sugar) molecule. They are joined in pairs to produce, usually, glucose or fructose. In some plants glucose and fructose are joined to produce sucrose, table sugar. This we harvest from sugarcane, sugar beets, or maple sap. The plant cell might use glucose to produce cellulose or package it as starch.

Notice that although one of the six PGAL molecules was removed from the cycle, five of them remained. This represents 15 carbon atoms in all. Through an ingenious series of reactions, these 15 carbons and their associated atoms are rearranged into three molecules of the very fivecarbon compound, ribulose phosphate, with which we started. Now this same ribulose phosphate is in a position to begin the process of CO2 fixation and eventual PGAL production once again.

In summary, the inputs required for the light-independent reactions are three molecules of $CO_{2,}$ hydrogen from the photolysis of water, and ATP. In the end the three carbons from the CO_2 can be accounted for by the harvest of a three-carbon (half-hexose) molecule. The remaining organic compounds are used to synthesize the pre-existing substances with which yet another three CO_2 molecules may combine once again.

The Four-Carbon Pathway

Most plants are referred to as C_3 plants because the first product of CO_2 fixation is a three-carbon compound, PGA. Some plants, referred to as C_4 plants, are able to fix carbon dioxide in the four-carbon compound known as *oxaloacetate*. After the oxaloacetate is produced, it moves to adjacent cells, where it is oxidised with the release of carbon dioxide. This carbon dioxide then reacts with ribulose bisphosphate and finally enters the Calvin cycle.

What is the advantage of the C_4 pathway? Its essential feature is that it concentrates carbon dioxide in the cells that carry on photosynthesis by the C_3 pathway. Carbon dioxide enters the leaf through tiny pores called stomata that open and close in response to such factors as water content and light intensity. When the stomata are closed, the supply of carbon dioxide is greatly diminished, and in C_3 plants photosynthesis is significantly slowed. However, the enzyme *phosphoenolpyruvate (PEP) carboxylase*, which catalyzes the formation of oxaloacetate, has a high affinity for carbon dioxide and binds it effectively even when carbon dioxide is at low concentration. Thus in C_4 plants photosynthesis, and ultimately growth, can continue efficiently even under certain adverse conditions. Many quick-growing and aggressive plants possess the C_4 pathway, among them sugarcane, corn, and crabgrass. In crop plants the C_4 pathway makes high yields possible. If this pathway could be incorporated into more of our crop plants by genetic manipulation, we might well be able to greatly increase food production in some parts of the world.

8

CELLULAR RESPIRATION

During photosynthesis light energy from the sun is converted into chemical energy stored within the bonds of organic molecules, such as carbohydrates. Most cells depend upon these fuel molecules for the continuous input of energy required to carry on life processes. Although some ATP is manufactured during the light-dependent reactions of photosynthesis, that ATP is used up during the light-independent reactions. Thus even plant cells must depend upon energy-releasing pathways that enzymatically break down organic molecules. Those pathways are collectively referred to as *cellular respiration*, the process by which a living cell breaks down a fuel molecule to carbon dioxide and water, releasing the energy trapped within its chemical bonds.

THE REACTIONS OF CELLULAR RESPIRATION

One of the principal pathways of cellular respiration is that of the common nutrient glucose. The overall reaction for the metabolism of glucose follows:

$$\underset{\text{Glucose}}{C_6H_{12}O_6} + \underset{\text{Oxygen}}{6O_2} \rightarrow \underset{\text{Carbon dioxide}}{6CO_2} + \underset{\text{Water}}{6H_2O} + \text{Energy}$$

The cellular oxidation of glucose to carbon dioxide and water cannot occur in a single reaction, for there are no enzymes that can catalyze the direct attack of oxygen on glucose. Instead, the oxidation of glucose occurs in a sequence of reactions that can be grouped into

Table 8.1: Summary of Cellular Respiration.

Phase	Summary	Needed materials	End products
Glycolysis (takes place in cytoplasm)	Series of about 10 reactions during which glucose is degraded to pyruvate; net profit of 2 ATPs; hydrogens released; can proceed anaerobically	Glucose, 2 ATPs	Pyruvate, ATP, hydrogen
Formation of acetyl CoA (takes place in mitochondria)	Pyruvate is degraded, and combined with CoA to form acetyl CoA; CO_2 is released	Pyruvate, CoA, NAD	Acetyl CoA, NAD—H, CO_2
Citric acid cycle (takes place in mitochondria)	Series of reactions in which fuel molecule (part of acetyl coenzyme A) is degraded to hydrogen and carbon dioxide; aerobic	Acetyl CoA, H_2O, hydrogen acceptors (e.g., NAD)	CO_2, H, reduced hydrogen acceptors, CoA
Electron transport and chemiosmosis (take place in mitochondria)	Chain of several electron transport molecules; hydrogens (or their electrons) are passed along chain; energy released is used to generate proton gradient across inner mitochondrial membrane; as protons move through ATP synthetase in membrane, ATP is synthesized; for each pair of hydrogens that enter chain, maximum of 3 ATPs can be synthesized; aerobic	Hydrogen, ADP, P_i, oxygen	ATP, water

four phases:

1. Glycolysis, the conversion of the 6-carbon glucose molecule to two three-carbon molecules of pyruvate[1] with the formation of two ATPs.
2. Formation of acetyl coenzyme A. Pyruvate is degraded to a twocarbon fuel molecule and combines with coenzyme A, forming acetyl coenzyme A; carbon dioxide is released.
3. The citric acid cycle, which converts acetyl coenzyme A to carbon dioxide and removes electrons and hydrogens.
4. The electron transport system and chemiosmotic phosphorylation. The hydrogens and electrons removed from the fuel molecule during the preceding phases are transferred along a chain of electron acceptor compounds. As the electrons are passed from one electron acceptor to another, a proton gradient is established, which sets the stage for ATP production by chemiosmosis.

Glycolysis

The sequence of reactions that convert glucose to pyruvate with the production of ATP is termed *glycolysis*. Each reaction is regulated by a specific enzyme, and there is a net gain of two ATP molecules. The reactions of this pathway take place in the cytoplasm, and these reactions can proceed anaerobically, that is, in the absence of oxygen.

In a highly simplified explanation of glycolysis we can divide the process into two major phases. In the first phase the glucose molecule is split to form two molecules of the three-carbon compound glyceraldehyde-3-phosphate, or *PGAL*. (Note that this is the same compound formed during the light-independent reactions of photosynthesis.) This transformation requires energy, so the cell must invest two molecules of ATP in order to initiate the oxidation of glucose. We may summarise this step of glycolysis as follows:

$$\underset{\text{6-carbon compound}}{\text{Glucose}} + 2\text{ATO} \rightarrow\rightarrow\rightarrow \underset{\text{3-carbon comound}}{2\text{PGAL}} + 2\text{ADP} + 2\text{P}$$

Several arrows are used to indicate that the equation summarises a sequence of several reactions.

In the second phase of glycolysis two hydrogen atoms are removed from each PGAL (oxidising the PGAL), and certain other atoms are rearranged so that each molecule of PGAL is transformed into a molecule of pyruvate. During these reactions enough chemical energy is released from the fuel molecule to produce four ATP molecules.

$$\underset{\text{3-carbon compound}}{2\text{PGAL}} + 4\text{ADP} + 4\text{P}_i \rightarrow\rightarrow\rightarrow\rightarrow \underset{\text{3-carbon compound}}{2\text{Pyruvate}} + 4\text{H} + 4\text{ATP}$$

Note that in the first phase of glycolysis two molecules of ATP are consumed, but in the second phase four molecules of ATP are produced. Thus glycolysis yields a net energy profit of two ATP molecules.

The two hydrogen atoms removed from each PGAL immediately combine with the hydrogen carrier molecule, NAD:

$$\underset{\text{Oxidised}}{\text{NAD}^+} + 2\text{H} \rightarrow \underset{\text{Reduced NAD}}{\text{NAD} - \text{H}} + \text{H}^+$$

The fate of these hydrogen atoms will be discussed with the electron transport system and with anaerobic pathways.

The reactions of glycolysis take place in the cytoplasm. Such necessary ingredients as ADP, NAD, and phosphates float freely in the cytoplasm and are used as needed. For a more detailed description of glycolysis, see Focus on Glycolysis.

Formation of Acetyl CoA

Pyruvate molecules produced during glycolysis move into the mitochondria, and essentially all subsequent reactions of cellular respiration take place within these tiny power plants. Each three-carbon molecule of pyruvate is degraded by the removal of a carbon-oxygen segment and two hydrogen atoms. The carbon-oxygen segment is removed as $C0_2$, which eventually is expired (by way of the lungs in complex animals). The pyruvate molecule is converted to a two-carbon acetyl group, which reacts with a large, complex coenzyme called *coenzyme A*. This results in the formation of a compound called acetyl *coenzyme A (CoA)*.

$$2\text{pyruvate} + 2\text{CoA} + 2\text{NAD}^+ \rightarrow 2\text{acetyl—CoA} + 2\text{NAD—H} + 2\text{H}^+ + 2\text{CO}_2$$

Coenzyme A is manufactured in the cell from one of the B vitamins, pantothenic acid. Fats and amino acids can also be degraded to acetyl CoA and can enter the respiratory pathway at this point. The six carbon atoms from the original glucose molecule may now be accounted for as follows: four of the carbons are located in the two acetyl CoA molecules, and the other two were released as carbon dioxide.

The Citric Acid Cycle

The *citric acid cycle* (also known as the *Krebs cycle* after Sir Hans Krebs, who worked it out) is the principal series of reactions in which hydrogen atoms are removed from fuel molecules and transferred to carrier molecules, such as NAD, for further processing in the electron

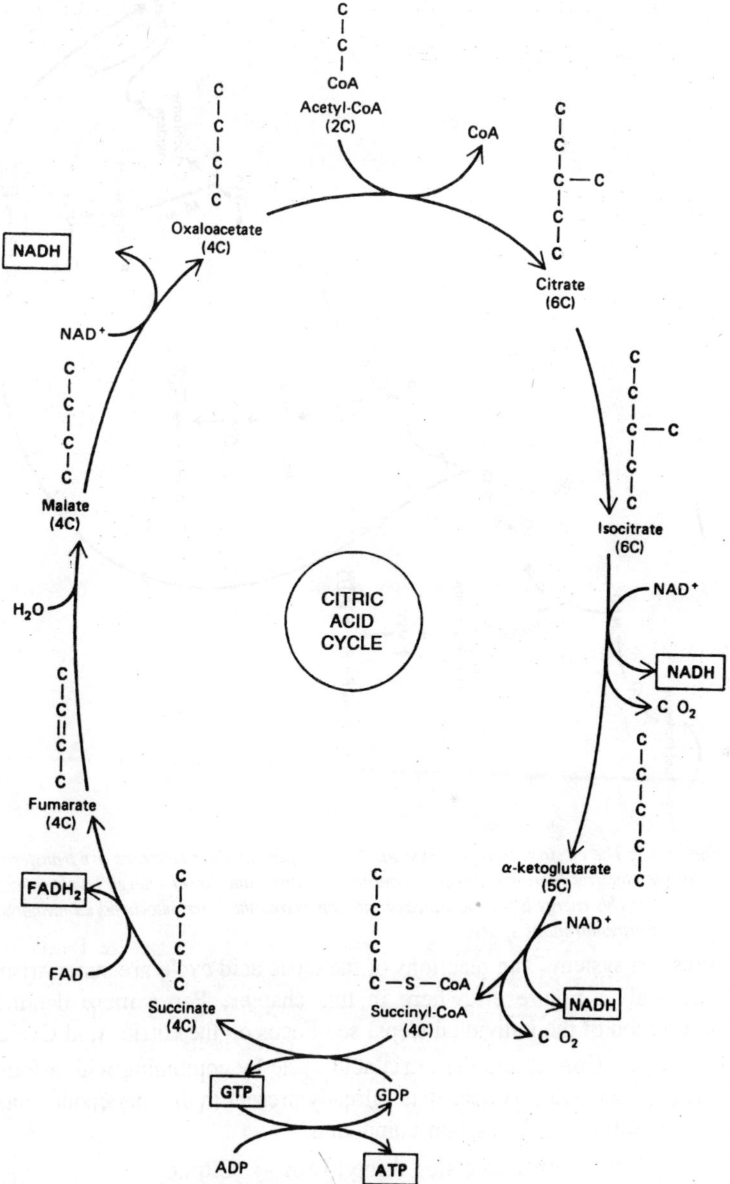

Figure 8.1: The citric acid cycle. During this series of reactions, acetyl coenzyme A, produced from glucose and other organic compounds, is metabolised to yield carbon dioxide and hydrogen. The hydrogen is immediately combined with NAD or FAD and is fed into the electron transport system.

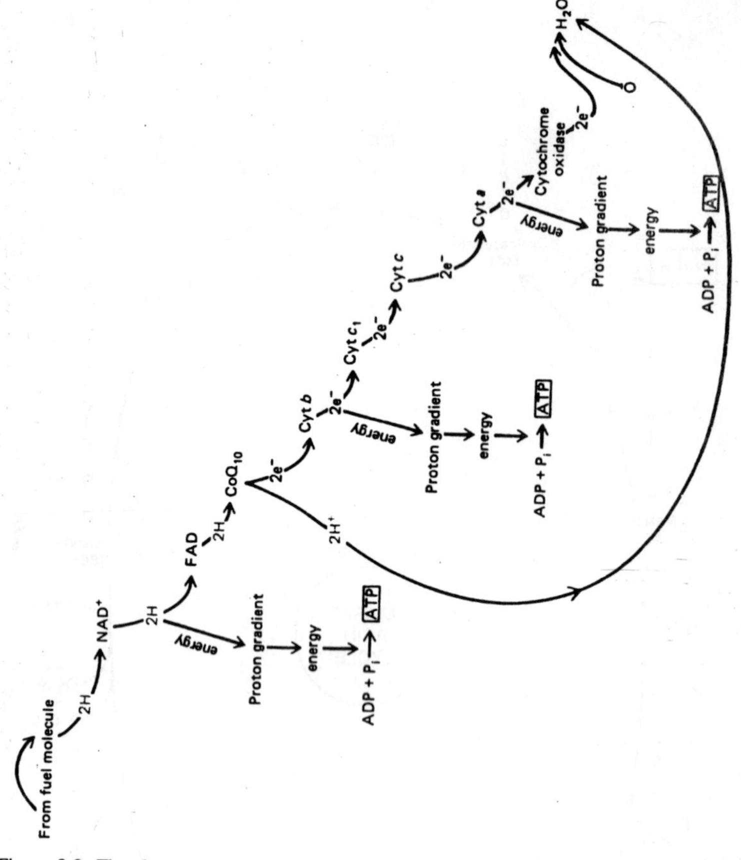

Figure 8.2: The electron transport system. As hydrogens or their electrons are transferred from one electron carrier molecule to another at lower and lower energy levels, energy is released. This energy is used to transport protons across the inner mitochondrial membrane to the intermembranous space.

transport system. The reactions of the citric acid cycle are summarised here and in Figure elsewhere in this chapter. For a more detailed description of the individual steps, see Focus on the Citric Acid Cycle.

Acetyl CoA enters the citric acid cycle by combining with a four-carbon compound, oxaloacetate (already present in the mitochondrion), to form citrate, a six-carbon compound.

Oxaloacetate+	Acetyl CoA →	Citrate
4-carbon compound	2-carbon compound	6-carbon compound

The citrate now goes through a series of chemical transformations so that it loses first one, then another carbon-oxygen segment in the form of CO_2. At the end of a complete cycle a four-carbon oxaloacetate is all that is left, and the cycle is ready for another turn. By this time the original pyruvate has lost all of its three carbons, or at least the equivalent, and may be regarded as having been completely consumed. Two turns of the cycle are necessary to process the two pyruvate molecules derived from the original glucose. Only one molecule of ATP is produced directly (by a substrate-level phosphorylation) with each turn of the citric acid cycle. How, then, is most of the ATP produced?

The Electron Transport System and Chemiosmotic Phosphorylation

Now let us consider the fate of all the hydrogens removed from the fuel molecule during glycolysis, acetyl CoA formation, and the reactions of the citric acid cycle. These hydrogens are first transferred to the hydrogen acceptors, NAD or FAD. NAD is the first hydrogen acceptor in the series of compounds that make up the electron transport system. As the hydrogens (or their electrons) are transferred from one to another of a series of some 15 different electron acceptor molecules, the hydrogen protons become separated from their electrons. The protons (H^+) are released into the surrounding medium and set up an electrochemical gradient across the inner mitochondrial membrane. This gradient provides the energy for ATP synthesis. This chemiosmotic mechanism is similar to the mechanism employed by the chloroplast to make ATP.

The electron acceptors of the electron transport chain are embedded within the inner membrane of the mitochondrion. Among the electron acceptors in the chain are *cytochromes*, a group of closely related proteins characterised by a central atom of iron. It is the iron that combines with the electrons from the hydrogen atoms. Cytochrome molecules accept only the electron from the hydrogen, rather than the entire atom. The last cytochrome in the chain passes the two electrons on to molecular oxygen. Simultaneously, the electrons reunite with protons. The chemical union of the hydrogen and oxygen produces water.

Oxygen is the final hydrogen acceptor in the electron transport system. This vital role of oxygen explains why we must breathe. What happens when the cells are deprived of oxygen? When no oxygen is available to accept the hydrogen, the last cytochrome in the chain is stuck with it.

When that occurs, each acceptor molecule in the chain may remain stuck with electrons, and the entire system may be blocked all the way back to NAD. No further ATPs can be produced by way of the electron

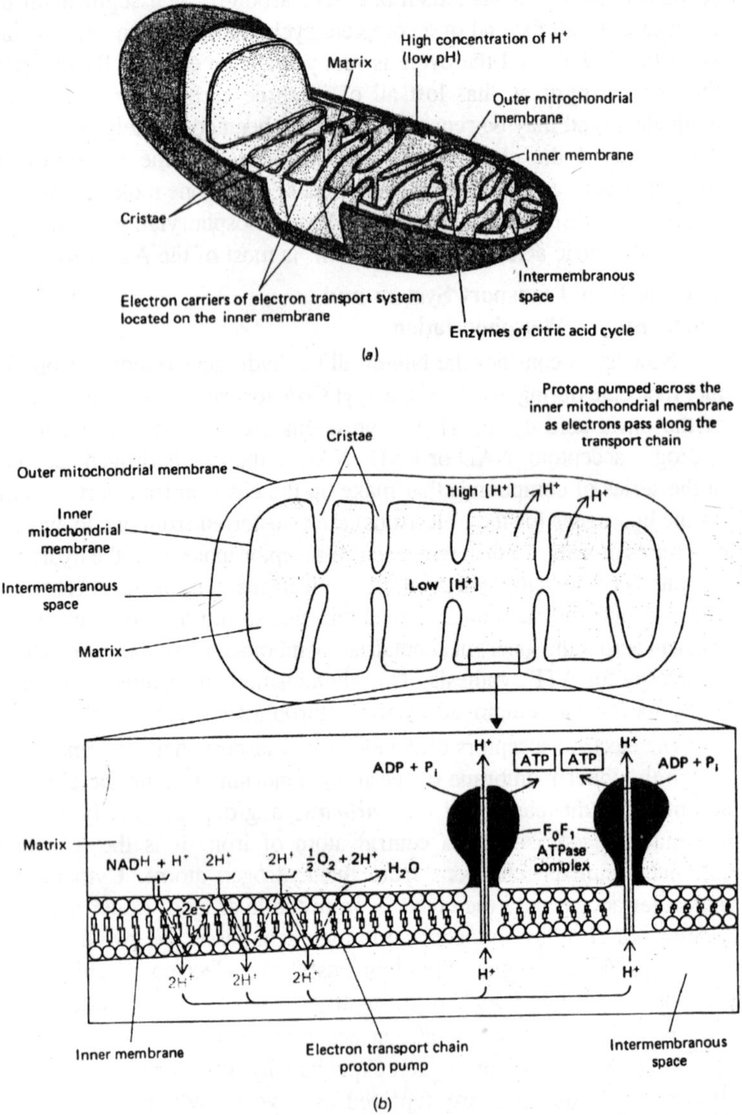

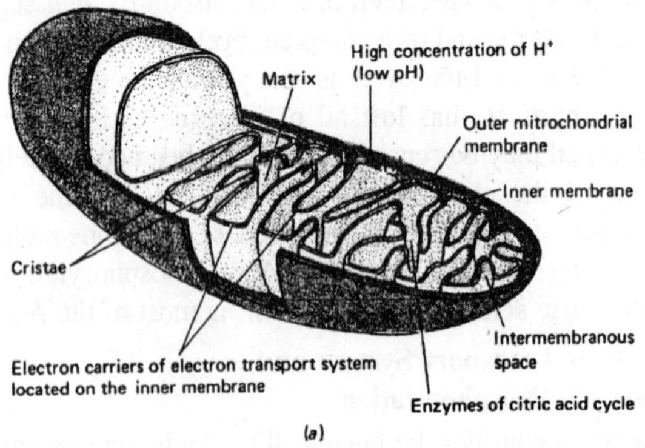

(a)

(b)

Figure 8.3: Chemiosmosis. (a) According to the chemiosmotic theory, the electron transport chain in the inner mitochondrial membrane is a proton pump. (b) The energy released during electron transport from substrate to oxygen is used to transport protons, H^+, from the mitochondria) matrix to the intermembranous space, where a high concentration of protons accumulates. There are three sites in the electron transport chain in which protons are transported.

transport system. Most cells of complex organisms cannot live long without oxygen because the amount of energy they can produce in its absence is insufficient to sustain life processes.

Lack of oxygen is not the only factor that may interfere with the electron transport system. Some poisons, including cyanide, inhibit the normal activity of the cytochrome system. Cyanide combines electrons tightly with one of the cytochrome molecules so that they cannot be passed along to the next acceptor in the chain.

As electrons are transferred along the acceptors in the electron transport chain, they are passed down an energy hill. Sufficient energy is released at three points to transport protons across the inner mitochondrial membrane, and ultimately to synthesize ATP. The passage of each pair of electrons from NADH to oxygen yields three ATPs.

Protons pumped outside the inner mitochondrial membrane can flow back to the matrix of the mitochondrion only through special channels in the inner membrane. These channels occur in the enzyme ATP synthetase, also known as the FO-F, complex. As the protons move down an energy gradient, the energy released is used to synthesize ATP. The ATP synthetase acts as a turbine converting one form of energy into another. The flow of electrons is tightly coupled to the phosphorylation process and will not proceed unless phosphorylation can also occur. Because the phosphorylation of ADP to form ATP is coupled with the oxidation of electron transport compounds, this entire process is referred to as *oxidative phosphorylation*.

REGULATION OF CELLULAR RESPIRATION

Cellular respiration requires a steady input of nutrient fuel molecules and oxygen. Under normal conditions these materials are adequately provided and do not affect the rate of respiration. Instead, the rate of cellular respiration is regulated by the amount of ADP and phosphate available. In a resting muscle cell, for example, ATP synthesis continues until all the ADP has been converted to ATP.

Then, when there are no more acceptors of phosphate, phosphorylation must stop. Since electron flow is tightly coupled to phosphorylation, the flow of electrons also stops. When an energy-requiring process like muscle contraction occurs, ATP is split to yield ADP and inorganic phosphate plus energy. The ADP formed can then serve as an acceptor of phosphate and energy to become ATP once again. Oxidative phosphorylation continues until all the ADP has again been converted to ATP. Because phosphorylation is tightly coupled to electron flow, the cell possesses a system of control that can regulate the rate of ATP production and adjust it to the momentary rate of energy utilization.

Although the reactions of glycolysis take place in the cytoplasm, those of the citric acid cycle and the electron transport system occur within the mitochondria. The outer membrane of the mitochondrion engages in active transport and is thought to regulate the intake of materials. Most respiratory enzymes are associated with the inner mitochondria) membrane.

It is thought that these enzymes must be exactly located with respect to one another so that hydrogens and electrons may be passed from one to another in the multitudinous and still somewhat bewildering metamorphoses of cellular respiration. No assembly line yet built can match the mitochondrial disassembly line for speed and efficiency. What a pity we cannot see it as it must be, vibrating with thousands of elaborate transformations every second, pouring forth streams of ATP, water, and carbon dioxide as it labors along with thousands like it in every one of the billions of cells of the body.

ANAEROBIC PATHWAYS

When sufficient oxygen is not available, the electron transport system and the citric acid cycle cannot operate, and metabolism may proceed *anaerobically*, that is without oxygen. Recall that the net profit of two ATPs produced during glycolysis does not require the presence of oxygen. When oxygen is present, the hydrogens removed from the fuel molecule during glycolysis pass through the electron transport system. In the absence of oxygen or of required enzymes this pathway is blocked, and the hydrogens must be disposed of in some other way.

Certain types of bacteria engage solely in anaerobic metabolism, a clear advantage to those that inhabit the soil or polluted waters where oxygen is in short supply. Other bacteria are able to shift to anaerobic pathways when deprived of oxygen. In *anaerobic respiration*, an inorganic compound, such as nitrate, serves as the final acceptor of electrons. Some of the bacteria important in recycling nitrogen carry on anaerobic respiration.

In *fermentation* an intermediate compound in carbohydrate metabolism is the final acceptor of electrons from NADH. Both anaerobic respiration and fermentation depend upon the reactions of glycolysis. When deprived of oxygen, yeast cells split carbon dioxide off from pyruvate. Hydrogen from NADH + H^+ is transferred to the resulting compound, forming *ethyl alcohol*, or drinking alcohol. Such anaerobic reactions are the basis for the production of beer, wine, and

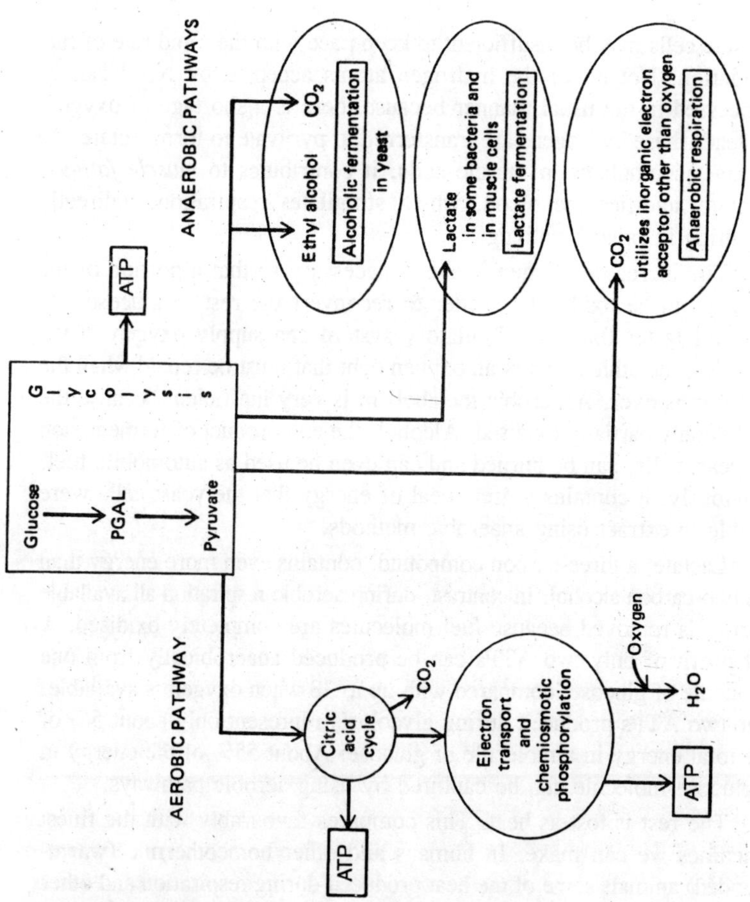

Figure 8.4: Comparison of anaerobic with aerobic pathways. When oxygen is not available, fermentation occurs in certain bacteria, yeast, and muscle cells.

other alcoholic beverages. Yeast cells are also used in the baking industry to produce the carbon dioxide that causes dough to rise.

An alternative pathway for hydrogens removed from the fuel molecule during glycolysis is to transfer them to the pyruvate molecule. When hydrogen atoms are added to pyruvate, *lactate*, the ionic form of lactic acid, is formed.

$$\text{Pyruvate} + 2\text{H} \rightarrow \text{Lactate}$$

Lactate is produced when bacteria sour milk or ferment cabbage to form sauerkraut. It is also produced during muscle activity in the muscle cells of humans and other complex animals. During strenuous physical activity, such as running, the amount of oxygen delivered to the muscle cells may be insufficient to keep pace with the rapid rate of fuel oxidation. Not all of the hydrogen atoms accepted by NAD can be processed in the usual manner because there is a shortage of oxygen. Instead, these hydrogens are transferred to pyruvate to form lactate. As lactate accumulates in muscle cells, it contributes to *muscle fatigue*. Lactate acidifies the blood, which stimulates respiration, indirectly resulting in rapid breathing.

The increased oxygen intake is necessary so that a portion of the lactate can be oxidised in order to reconvert the rest to glucose. By running faster than the circulatory system can supply oxygen to the muscles, the athlete incurs an oxygen debt that must be repaid when the exertion is over. Anaerobic metabolism is very inefficient because the fuel is only partially oxidised. Alcohol, the end product of fermentation by yeast cells, can be burned and can even be used as automobile fuel. Obviously, it contains a great deal of energy that the yeast cells were unable to extract using anaerobic methods.

Lactate, a three-carbon compound, contains even more energy than the two-carbon alcohol. In contrast, during aerobic respiration all available energy is removed because fuel molecules are completely oxidised. A net profit of only two ATPs can be produced anaerobically from one molecule of glucose, compared with up to 38 when oxygen is available. The two ATPs produced during glycolysis represent only about 5% of the total energy in a molecule of glucose. About 55% of the energy in a glucose molecule can be captured by using aerobic pathways.

The rest is lost as heat. This compares favourably with the finest machines we can make. In humans and other homeothermic (warm-blooded) animals some of the heat produced during respiration and other metabolic activities is utilised to maintain a constant body temperature.

Table 8.2: Comparison of Aerobic and Anaerobic Pathways.

	Anaerobic	Aerobic
Maximum ATP profit	2 ATPs	36-38 ATPs
Final hydrogen acceptor	Pyruvate	Molecular oxygen
End products	Ethyl alcohol in yeast, certain bacteria, lactate in muscle cells	CO_2, H_2O

The inefficiency of anaerobic metabolism necessitates a large supply of fuel. By rapidly degrading many fuel molecules a cell can compensate somewhat for the small amount of energy that can be gained from each. To perform the same amount of work, an anaerobic cell must consume up to 20 times as much glucose or other carbohydrate as a cell metabolizing aerobically. Skeletal muscle cells, which often metabolize anaerobically for short periods, must therefore store large quantities of glucose in the osmotically inactive form of glycogen.

9

CELL MOTILITY

All organisms depend on movements of molecules from place to place. For example, ATP made in one region is often used elsewhere in the cell, and nutrients must be captured from the environment. Small organisms have no problem in randomising their internal molecules since diffusion is a fast process over short distances like the few micrometers that span a bacterium.

However, large cells need to mix their components, and they have developed special internal machinery to get molecules and ions from one place to another. This machinery is also used in big cells and multicellular organisms to position and move organelles and change cell shape. In addition, many organisms have evolved life-styles that depend on their ability to move around. Such mobility can be useful in feeding, in avoiding becoming food, and in sexual reproduction. Thus, natural selection has favoured the evolution of motile systems.

Cells have evolved a variety of ways of moving. In some cases, movement is only intracellular, consisting of the active displacement or propulsion of materials and structures inside the cell. A common example of this is rapid *cytoplasmic streaming*, seen prominently in some algae and plant cells, in which it is called *cyclosis*. Another example is the movement of chromosomes during mitosis and meiosis. Other forms of intracellular movement are the fast propulsion of vesicles in the axon of a nerve cell and the rapid shuttling of pigmented granules in some kinds of pigmentcontaining cells.

In addition to strictly intracellular movement, many cells are capable of various forms of motility in which the cell changes its shape or moves in relation to its environment. An example is the gliding movement of certain species of bacteria and diatoms. The pinching in two of a cell during cytokinesis is another type of movement. Some cells, such as amoebae, move by cytoplasmic streaming (*amoeboid movement*).

Some cells, like bacteria, sperm, some protozoa, and algae, move by means of flagella. Still others, including some protozoa and cells in many kinds of larvae, are propelled by cilia. Many kinds of bacteria are propelled by flagella, but the structure and operation of bacterial flagella are very different from the structure and operation of eukaryotic flagella. Certain animal tissue cells possess cilia (e.g., the ciliated epithelium of the respiratory tract), but since these ciliated cells are held stationary in the tissue, the motion of their cilia causes the external medium to flow across their surface.

Cellular movement is developed to its highest degree in muscle cells. Since muscle cells remain in relatively fixed positions, their contraction produces displacement of other parts of the organism, for example, the movement of bony parts (*skeletal muscle*), propulsion of blood (*cardiac muscle*), and the propulsive movements, *peristalsis*, of the alimentary tract (*smooth muscle*).

Movement of all cells is achieved by a mechanochemical process in which chemical energy is used to do mechanical work moving a cell or cell parts. In prokaryotes two kinds of movement that probably involve different mechanochemical processes are known: (1) gliding movements in some species of bacteria, and (2) flagellar movement.

In eukaryotes, three mechanochemical processes are known. Each involves the interaction of two kinds of proteins that produces force by hydrolyzing ATP to ADP and P_i. The three known processes in eukaryotes, which account for most kinds of cell movement, are the following:

1. Interaction of the protein actin with myosin. The actin-myosin system is the basis of all muscle contraction in invertebrates and vertebrates. It is also the mechanism for generating cyclosis in plant and algal cells, and cytokinesis in animal cells and most protozoa; it is probably responsible for cytoplasmic streaming in amoeboid movement.
2. Interaction of the proteins tubulin and *dynein*. The tubulin-dynein system is the basis of all ciliary and flagellar movement in eukaryotic cells.
3. Interaction of the proteins tubulin and *kinesin*. The tubulin-kinesin system is the basis for the fast transport of vesicles and probably other particulate structures, for example, transport of vesicles in nerve axons and other kinds of cells.

MOVEMENT BASED ON ACTIN-MYOSIN INTERACTION

Of all mechanisms of movement the interaction of actin and myosin in the contraction of muscle is the best understood. It is useful to begin discussion of cell movement with muscle contraction because knowledge of how it works can serve as a basis for understanding other forms of movement.

Structure and Function of Muscle Cells

Muscle cells contain large amounts of a few kinds of proteins that are highly ordered to form contractile structures. In skeletal and cardiac muscle, contractile structures are ordered in a way that gives rise to characteristic cross striations. The ordering of contractile elements in smooth or ***involuntary muscle*** does not produce cross striations. The molecular mechanisms of contraction of smooth muscle cells is fundamentally similar to that in striated muscle. However, the organization of the contractile machinery in smooth muscle is less well understood than for striated muscle, so only the striated muscle cell is discussed here.

Structure of a striated muscle cell

In the muscles of vertebrates the individual cells of skeletal muscle are multinucleated cylinders. They vary in diameter from 10 to 100 μm and vary in length from a few millimeters to several centimeters. These large cells are formed during muscle development by the fusion of many reproducing, uninuclear cells called *myoblasts*. Shortly before fusion, myoblasts stop reproducing and enter G0, where they remain permanently. Several thousand myoblasts may fuse to form a single, mature multinucleated muscle cell. Many thousands of multinucleated muscle cells are, in turn, packed together in bundles to form a muscle. At the two ends of the muscle, the muscle cells are attached-to the collagenous fibers of tendons, and the tendons are usually attached to bones.

Myoblasts and the multinucleated cells produced by fusion of myoblasts synthesize enormous quantities of the few kinds of proteins that make up the contractile machinery. These proteins become organised in fibrous structures that give rise to the cross striations seen in a mature muscle cell. The many sets of cross striations in a single muscle cell reflect the presence of repeating contractile units called *sarcomeres*. An individual sarcomere occupies about 2.5 μm or 0.01 percent of the length of a cell, and hence one muscle cell may contain many thousands of sarcomeres. Each sarcomere contains a set of several kinds of cross striations that are visible by light microscopy but more effectively studied by electron microscopy. The cross striations consist of various bands and lines and are the result of a highly ordered arrangement of the proteins that make up the contractile machinery.

At each end of a sarcomere is a cross striation called the *Z line*, which forms the boundary shared by two successive sarcomeres. About 1.6 μm of the central portion of a sarcomere is occupied by a more densely staining structure called the A band. Between the end of an A band and the Z line is a less dense region called the I band. A single I band occupies portions of adjacent sarcomeres and is transected by a Z line. In short, the muscle cell consists of alternating A bands and I

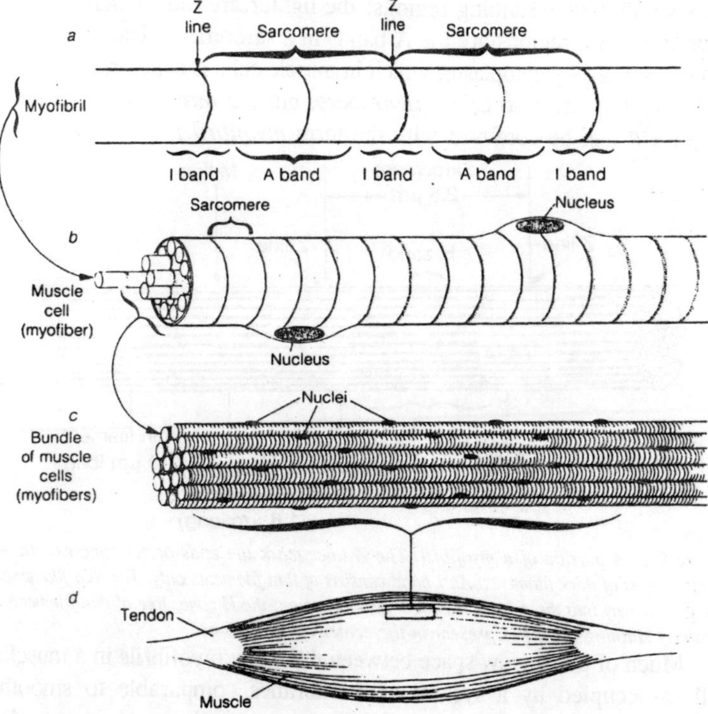

Figure 9.1: Structure of skeletal muscle. (a) A myofibril is a long cylinder consisting of tandemly arranged sarcomeres. A sarcomere consists of a darkly staining central region, the A band and two adjacent lightly staining regions, the I bands. Very darkly staining Z lines mark the borders of successive sarcomeres. (b) Many myofibrils are packed inside one muscle cell (myofiber). The myofibrils are in parallel register and impart the sarcomere banding pattern to the muscle cell. (c) A bundle of muscle cells. Muscle cells are multinucleated. (d) Bundles of muscle cells are in turn arranged into larger bundles within a muscle.

bands organised into sarcomeres, and each sarcomere consists of a central A band flanked on each end by one-half of an I band. The A band is marked through its center by a less dense region called the H zone, which contains a central line called the M line. The significance of these structural arrangements will become clear a little later in the discussion of the arrangement of the molecules that bring about contraction.

The structural basis of cross striations is found in the many *myofibrils* that run lengthwise in a muscle cell. The myofibrils are constructed from contractile proteins. They are all parallel and in register, so their cross striations are lined up with one another. A myofibril consists of lighter and darker staining regions; the lighter are the I bands and the darker staining regions are the A bands in a sarcomere. The myofibrils are surrounded by cytoplasm, which in muscle cells is called *sarcoplasm*. (Muscle cells are also called myofibers, but the term is avoided here because it can be confused with the term *myofibril.)*

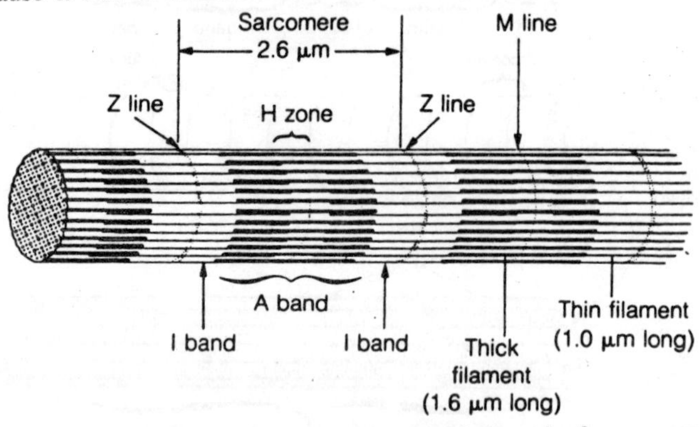

Figure 9.2: A portion of a myofibril. The Z lines mark the ends of sarcomeres. An A band consists of thick filaments. An I band consists of thin filaments only. The thin filaments extend part way into the A band leaving a central region, the H zone, free of thin filaments. A faintly staining M line is present in the center of the H zone.

Much of the narrow space between the many myofibrils in a muscle cell is occupied by a system of membranes comparable to smooth endoplasmic reticulum, but which has a particular role in muscle contraction; it is called the *sarcoplasmic reticulum*. The many nuclei occupy the outer region of the cylindrically shaped cell, positioned just inside the plasma membrane, which in muscle cells is called the *sarcolemma*. Organelles such as the Golgi complex, rough endoplasmic reticulum, free ribosomes and polysomes are scanty and lie concentrated around nuclei.

Glycogen particles scattered through the cytoplasm provide a store of carbohydrate for glycolysis and oxidative phosphorylation. Muscle cells are richly supplied with glycolytic enzymes and possess numerous mitochondria scattered among and closely adjacent to the myofibrils, which allows for efficient transfer to myofibrils of the large amount of ATP needed to power contraction.

Each of the many myofibrils in a muscle cell is made up of a number of filaments called *myofilaments*. There are both thick and thin myofilaments.

Thick myofilaments in mammals are 1.6 μm long and are arranged in a single bundle with all of their ends in precise register. The bundle of thick myofilaments in a myofibril is what gives rise to the dark staining A band. The thin myofilaments are about 1.0 μm long and are arranged in two bundles, one extending into the sarcomere from each end, where they are attached to the Z lines.

The thin myofilaments traverse one-half of an I band and enter the A band. In the A band the thin myofilaments interdigitate with the thick filaments and extend up to the edge of the H zone near the middle of the sarcomere. The two borders of the H zone are defined by the terminations of the two bundles of thin myofilaments that extend in from the two Z lines of the sarcomere.

Thus, in a cross-sectional view of the sarcomere through an I band only thin myofilaments are visible. Through part of an A band, both thick and thin myofilaments are seen, but in the H zone only thick myofilaments are visible.

The thick and thin myofilaments interdigitate in the A band in a highly ordered way in vertebrate striated muscle. Each thick myofilament is surrounded by six thin myofilaments.

Structural changes in a sarcomere during contraction

During maximum contraction of a skeletal muscle, each cell shortens about 40 percent. This shortening is accomplished by a uniform reduction in length of all the sarcomeres in the cell (from 2.6 μm to 1.7 μm). Shortening of a sarcomere brings the two Z lines closer together, without changing the length of the central A band (composed of thick myofilaments), but almost all of the H zone and the two I bands disappear. The disappearance of the H zone and I bands occurs because during contraction the thin filaments move farther into the A band. In full contraction the ends of the thin myofilaments extend in from the two Z lines and almost meet.

Electron microscopic observations on thin and thick myofilaments

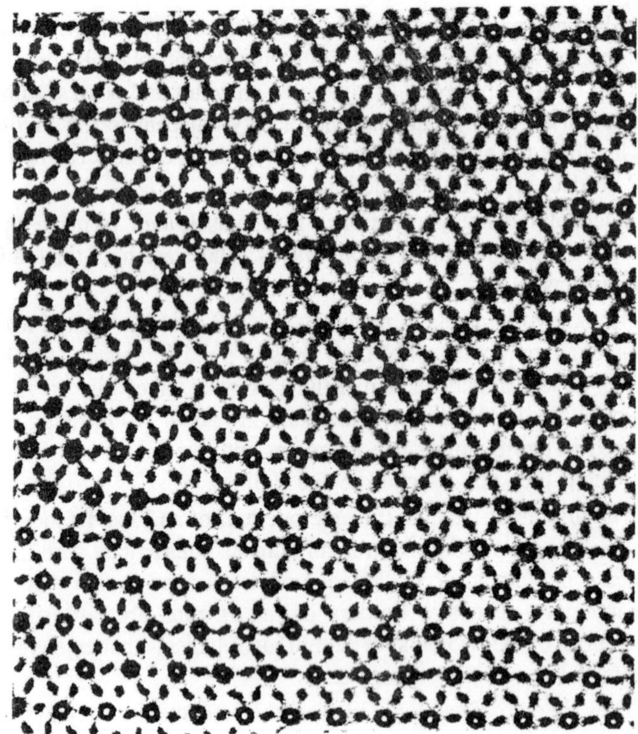

in relaxed and contracted muscle have revealed that muscle contraction occurs by the sliding of the thin myofilaments along the thick myofilaments. Thin and thick myofilaments interact with one another in a way that causes the thin myofilaments to be drawn into or to slide into the A band. The farther the thin myofilaments slide inward along the thick myofilaments, the greater the contraction.

Structure of thick myofilaments

Analysis of the proteins that make up the thin and thick myofilaments in a myofibril has enhanced understanding of the mechanisms by which the myofilaments are caused to slide along one another. When striated muscle cells are soaked in a physiological saline solution containing detergent to solubilize plasma membranes, the water-soluble proteins of the sarcoplasm are removed. The proteins that make up the myofibrils are not soluble in this solution; therefore, the sarcomeric structure of cells remains intact. If the muscle is now soaked in a cold alkaline solution of KCl (0.6 M), 50 percent of the protein is extracted, and the thick myofilaments disappear. Analysis of the extracted protein shows that thick myofilaments are composed almost entirely of a single type of protein, namely, myosin.

Myosin is a very large protein (molecular weight, 500,000) consisting of six polypeptide chains, two large ones and four smaller ones. The two large polypeptides are identical and have molecular weights of 200,000. They each contain about 1800 amino acids, which places them among the longer polypeptides known.

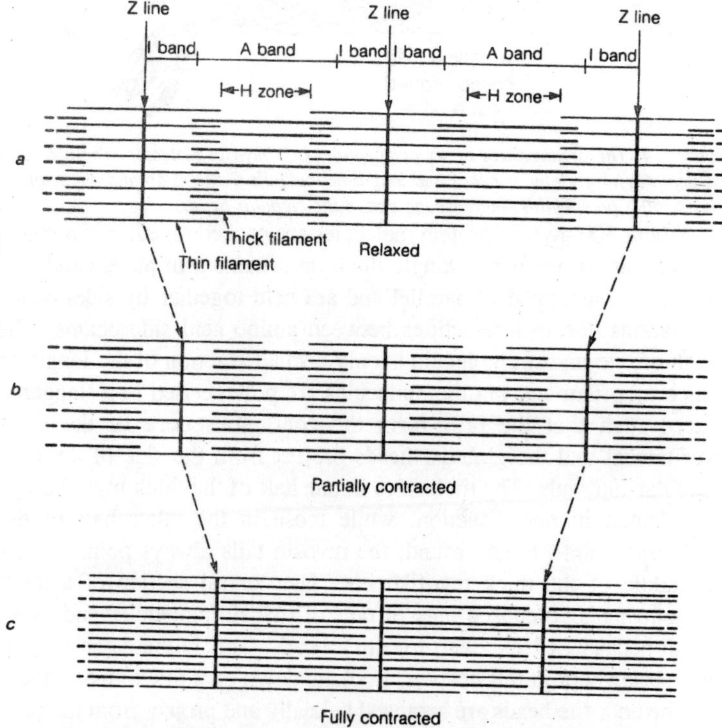

Figure 9.3: Contraction of muscle. (a) Relaxed myofibril. The I bands are at their longest and the Z lines are at maximum separation. The H zone in the A band is long. (b) Partially contracted myofibril. The thin microfilaments have been drawn further into the A band, resulting in shortening of I band and the H zone. The Z lines are now closer together. (c) Fully contracted myofibril. The I band has almost completely disappeared as a result of thin filaments being drawn into the A bands. The ends of thin filaments almost meet in the center of the sarcomere, and the H zone is barely visible.

A large portion of these chains is coiled into α-helices. In myosin the α-helical portions of the two polypeptides are wound around each other, forming a long linear molecule or tail. The remaining part of each chain is folded into a globular configuration that forms a head portion of this large polypeptide. The four small polypeptides (molecular weight from 16,000 to 21,000) are bound in pairs to the two globular portions of the large polypeptides, thus enlarging the globular head of a myosin molecule.

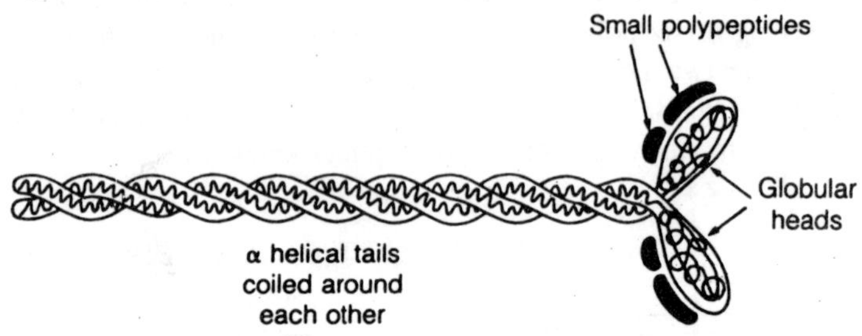

Figure 9.4: The arrangement of six polypeptides in a myosin molecule. The two large polypeptides have globular heads and long a-helical tails that coil around one another. Four smaller polypeptides increase the size of the globular heads.

About 500 myosin protein molecules are packed together in a highly ordered pattern to form a single thick myofilament in an A band. All the tails are arranged in parallel and are held together by side-to-side associations due to interactions between amino acid side chains. The length of one myosin molecule is only a small fraction of the length of a thick myofilament, but the molecules are polymerised in a staggered pattern such that the tails form the continuous core of the thick myofilament and the globular heads project from the side of the core at regular intervals. The molecules in one half of the thick myofilament are oriented in one direction, while those in the other half of the myofilament are turned around; the myosin tails always point toward the middle of the thick myofilament. The central section of a thick myofilament is composed only of tails of myosin molecules and looks smooth because it lacks the projecting heads. This is the clear zone in Figure elsewhere in this chapter. Looking down on a thick myofilament from the end, the heads are arranged helically and project from the core

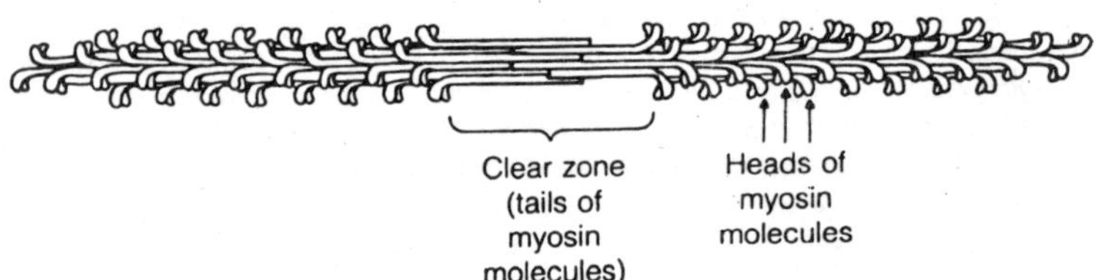

Figure 9.5: Structure of a thick filament. (a) Schematic diagram of myosin molecules aggregated to form a thick filament, with globular heads projecting outward. (b) Electron micrograph of a myosin thick filament isolated from scallop muscle. The clear zone in the center is in the H zone in a sarcomere.

in six directions. This results in six lines of heads that are in positions to interact with the six thin myofilaments that surround each thick myofilament in a myofibril. These interactions between heads and thin myofilaments are, as we shall see, the crux of the sliding filament mechanism described later. Furthermore, the head of the myosin molecule, in addition to its interaction with thin myofilaments, possesses an active enzymatic site that splits ATP into ADP + P_i. This splitting of ATP provides energy to fuel muscle contraction. The energetics of contraction are discussed later.

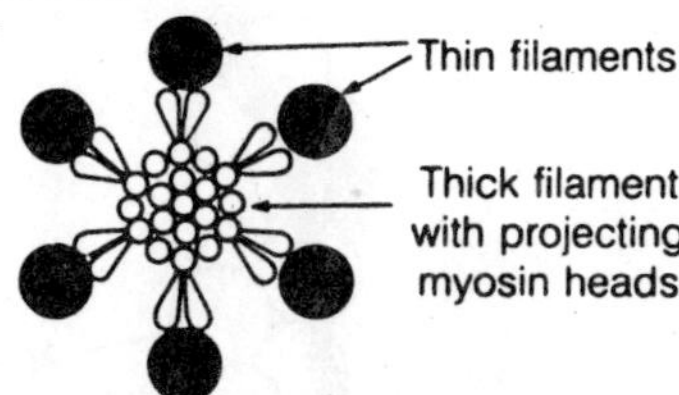

Figure 9.6: Schematic diagram of a thick filament seen end on. Pairs of heads of myosin molecules project radially toward six surrounding thin filaments.

Myosin makes up at least 95 percent of the mass of thick filaments. Other proteins are present in thick myofilaments in trace amounts. One of the trace proteins, named the *C-protein*, is stretched out in parallel with the myosin tails, winding around the core of the thick myofilament. This arrangement suggests that Cprotein serves to hold the myosin tails together, stabilizing the core of the thick myofilament.

Another protein present in small amounts is bound to the very center of the thick myofilament, where only myosin tails are present, and gives rise to the M *line* in the middle of the sarcomere. The function of this protein is not known, but it may have a role in connecting the thick myofilaments with one another and maintaining their precise order.

Structure of thin myofilaments

When muscle from which myosin has been extracted (with alkaline 0.6 M KCl) is subsequently soaked in cold, slightly alkaline water, the major proteins of the thin myofilaments dissolve. The principal protein in this extract is actin, which was described in the discussion of the cytoskeleton in other chapter of this book. When actin is solubilised from the thin myofilament, it is recovered in individual globular molecules called G-actin. When the inorganic ion concentration of a solution of pure G-actin is raised to the level present in intact muscle, the globular molecules join together to form long chains called F-actin. This is the form in which actin occurs in the thin myofilaments. In the thin myofilament, two chains of F-actin are wound around each other in helical fashion as described in other chapter of this book. The double helix has a diameter of about 9.5 run, which is the diameter of a thin filament.

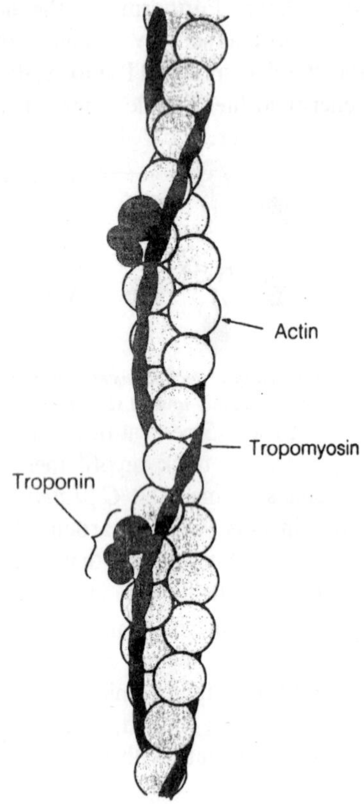

Figure 9.7: A model of a thin filament. Two tropomyosin (light pink) filaments, eah composed of two subfilaments, win around the actin chain. They block the binding of teh globyular heads of myosin molecules (in thich filaments) to the actin molecules Troponin consists of three polypeptides and binds both to actin and to tropomyosin.

Actin is the major component of thin myofilaments, but two other proteins, *tropomyosin* and *troponin*, are also present. The polypeptides of tropomyosin have an α-helical secondary structure, and are stretched out as long, thin molecules attached lengthwise to the F-actin. Purified tropomyosin molecules.in solution stick also to each other in end-to-end fashion. This end-to-end association gives rise to long filaments that lie in the two grooves of the F-actin double helix. Each tropomyosin molecule is about 40 run long and spans seven actin molecules.

The third protein of thin filaments, troponin, is a globular molecule consisting of three polypeptides. One molecule of troponin is bound to each molecule of tropomyosin. As described below, tropomyosin and troponin regulate contraction by regulating the interaction of thin and thick filaments, that is, the interaction of F-actin with myosin.

The thin myofilaments are all anchored at one end by strong attachment to a Z line. The Z line consists of a protein called a-actinin, which forms a multimolecular plate across the sarcomere (the Z line) that appears to bind the ends of F-actin chains (thin filaments). The aactinin forms the connections between the thin myofilaments of one sarcomere and those in the next. It therefore plays an essential mechanical role in linking together the myofibril. The polarity of F-actin chains was noted previously in other chapter; F-actin chains are elongated by addition of G-actin monomers; the addition occurring far more readily at one end than the other.

The polar nature of F-actin can also be demonstrated directly by the binding of modified myosin. Myosin molecules are digested briefly with the protease papain, which cuts them at the base of the double head, releasing the tail and the two halves of the head. When the purified heads are added to actin microfilaments, they bind in a highly regular pattern in which the myosin heads project from the microfilament at an angle.

When observed by electron microscopy a microfilament decorated with myosin heads gives the impression of a chevron or a line of arrowheads, with the barbs of the arrowheads, formed by the myosin heads, all facing in the same direction. With this method all the microfilaments in the I band on one side of a Z line have the same orientation; the arrow tips point away from the Z line and the barbs are directed toward the Z line. During the formation of microfilaments, G-actin monomers are added more rapidly to the end corresponding to the direction of the barbs, that is, the end at the Z line.

The interaction of actin with myosin in muscle contraction

Electron microscopic observations have indicated that muscle cells contract by an interaction of thick and thin filaments that causes the thin filaments to be drawn farther into the A band. By observing the interactions of the various purified muscle proteins, we have achieved considerable understanding of the molecular basis of the filament sliding.

When pure actin and pure myosin are mixed in solution, they form filamentous complexes called *actomyosin*. Actomyosin filaments can be

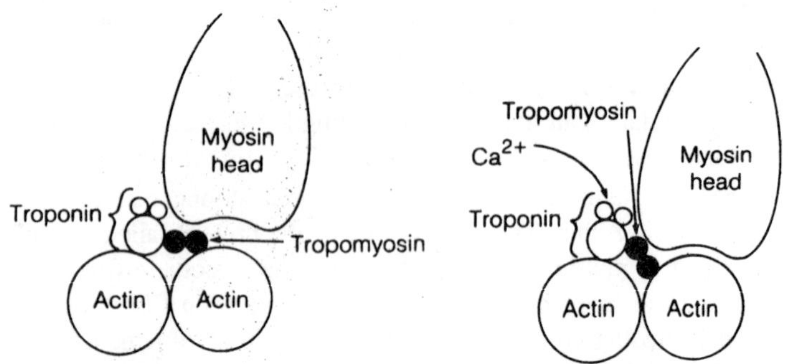

Figure 9.8: Schematic diagram of the interaction of myosin and actin. (a) In the absence of Ca^{2+} tropomyosin (red) prevents the binding of myosin heads to actin. (b) When the Ca^{2+} concentration rises, Ca^{2+} binds to a subunit of troponin, which causes the tropomyosin to shift slightly into the groove of the actin thin filament. The shift in position of tropomyosin allows the myosin heads to bind to actin. Lowering of the Ca ^{21} concentration results in reversal of these events.

made to aggregate into microscopically visible threads. One way to accomplish this is to layer actomyosin on the surface of water. The film of actomyosin can be skimmed from the water surface to make a fiber or thread, much as a film of scum from boiled milk can be collected into a threadlike mass.

When such artifically prepared actomyosin threads are immersed in a solution of ATP, they contract. With an appropriate arrangement, actomyosin threads can be made to lift small weights when exposed to ATP. These early experiments demonstrated that muscle contraction is the result of an interaction between actin and myosin that is modified by ATP.

Roles of Ca^{2+} tropomyosin, and troponin in contraction

Ca^{2+} ions play a central role in regulating muscle contraction. When Ca^{2+} ions are injected into a muscle cell, it contracts. Also, direct evidence for intracellular release of Ca^{2+} ions associated with contraction has come from injection of *aequorin*, a protein obtained from a luminescent jellyfish. Aequorin emits photons (*bioluminescence*) when it binds Ca $^{2+}$ ions. Aequorin has been injected into muscle cells and the muscle then stimulated to contract.

A very brief burst of light is seen being emitted for a fraction of a second *before* the contraction. By reconstitution experiments researchers have been able to learn about the roles of Ca^{2+} ions, tropomyosin, and troponin in regulating muscle contraction. Actomyosin threads prepared from pure actin and pure myosin contract whether or not Ca^{2+} ions are present in the solution. When tropomyosin and troponin are added to actomyosin, ATP induces contraction only when Ca $^{2+}$ ions are present. Thus, tropomyosin and troponin prevent contraction, but Ca^{2+} ions reverse this inhibition. It is now known that in the absence of Ca^{2+} ions, tropomyosin, which is arranged along the grooves of the F-actin double helix, blocks the interaction of F-actin with myosin and therefore blocks contraction. Troponin, which is bound to the tropomyosin, has a Ca^{2+} ion binding site. When

added to the system, Ca^{2+} ions become bound to troponin. The binding of Ca^{2+} ions causes a change in the shape of the troponin molecule. However, the troponin molecule is bound to tropomyosin, and the Ca^{2+} ioninduced change in troponin shape causes the tropomyosin to move deeper into the grooves of F-actin. This shift in the location of tropomyosin uncovers sites on the F-actin chains to which myosin can now bind. The binding of actin and myosin allows contraction to take place.

The interaction between myosin and actin occurs between the globular head portion of myosin and actin molecules. By electron microscopy this interaction is observed as cross bridges between thick and thin filaments. The cross bridges, which correspond to the heads of myosin molecules, connect the thick and thin filaments only when the muscle is contracting. In the relaxed state the cross bridges are released because the interaction of the heads of myosin molecules with the F-actin of the thin filaments is blocked by tropomyosin. Indeed, not only is the interaction blocked in relaxed muscle, but the myosin heads also appear to be retracted closer to the thick filament.

Molecular mechanism of muscle contraction

The studies of intact cells by electron microscopy, physiological analyses of living muscle cells, and biochemical studies on isolated and recombined components have all contributed to a rather complete picture of muscle contraction.

Myosin is organised into thick filaments in the A band, with the heads of molecules projecting at precise spacings from the axis of the thick filament. The myosin heads bind ATP. Normally, the concentration of ATP (the energy charge) in the sarcoplasm is high, and in relaxed muscle the binding sites on myosin are occupied by ATP, which is cleaved into ADP and P_i by the ATPase of myosin heads as soon as they are bound. With ADP and P_i bound to them the myosin heads are now poised to interact with actin and bring about contraction, but the interaction is blocked by tropomyosin. The interaction is brought about by Ca $^{2+}$ ions.

In relaxed muscle the concentration of Ca^{2+} ions is insufficient to bring about Ca^{2+} ion binding to the troponin of thin filaments. Under these circumstances, the tropomyosin of thin filaments is located slightly out of the groove of the F-actin chains in a position that blocks the interaction of myosin heads with F-actin. The arrival of a nerve impulse at the muscle cell causes an extremely rapid flux of Ca^{2+} ions into the sarcoplasm, and the myofibrils are bathed with these ions (control of the Ca^{2+} ion concentration in the sarcoplasm is discussed below).

With the sudden increase in Call ion concentration in the sarcoplasm, the Ca^{2+} ion-binding sites of troponin become saturated. The configurational change in troponin molecules induced by Ca^{2+} ion-binding draws the tropomyosin molecules deeper into the grooves of the F-actin of the thin filaments, thereby unmasking the sites on F-actin to which myosin heads can bind.

Myosin heads, each of which is already in an activated state by virtue of having bound ADP + P_i to it, bind to F-actin and form cross bridges between thick and thin filaments. The binding of myosin to actin induces the release of ADP and P_i from the myosin, causing the myosin heads to change their structure and revert to the nonactivated state.

The conversion of activated myosin to nonactivated myosin is associated with a change in molecular configuration, and this change somehow exerts a force on the thin filaments, pulling them a short distance further into the A band (about 10 nm per cross bridge). ATP can then bind once more to the empty site on the myosin molecule, and this causes myosin to fall off the actin, releasing the cross bridge, reexciting the myosin, and reextending it to a configuration in which it can cleave the ATP to ADP and P_i.

The myosin is ready once again to bind to the thin filament and give it another pull. The action of multiple, asynchronous bridge cycles achieves a smooth, extensive sliding of the thin filaments into the A band, causing the sarcomere to shorten and thereby producing muscle contraction. This somewhat complicated chain of events, extending from the time of arrival of a nerve impulse at the muscle cell to the actual contraction of the muscle, takes place in a small fraction of a second. Researchers are still trying to determine how myosin accomplishes its pulling action on the thin filaments.

As long as nerve impulses continue to arrive at a muscle cell, the concentration of Ca^{2+} ions in the sarcoplasm remains high, and interaction of thick and thin filaments can continue. Thus, as rapidly as myosin heads revert to a nonactivated state, they are reactivated by binding of

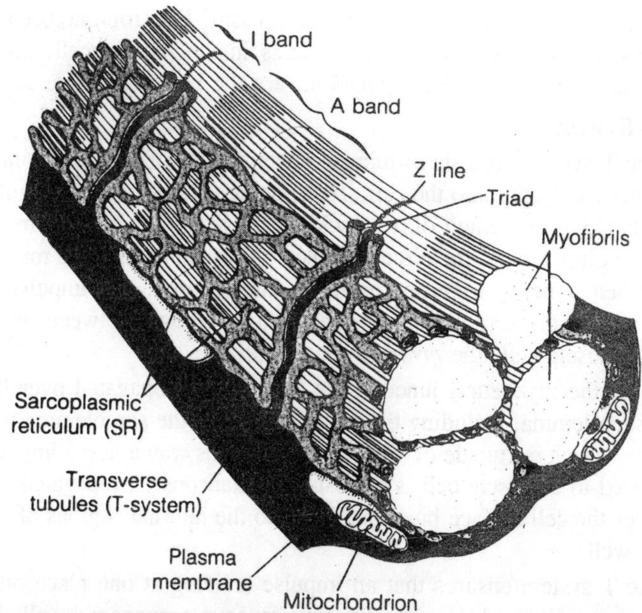

Figure 9.9: A drawing of part of a skeletal muscle cell showing the T system invaginations of the sarcolemma (plasma membrane) form a system of transverse tubules that relay a contraction signal from a neuron to the sacroplasmic reticulum in the interior of a cell.

ATP molecules and again interact with F-actin to exert another pull. Repeated cycles of myosin interaction with actin, occurring in a small fraction of a second, results in a tendency for every sarcomere in a stimulated muscle cell to shorten.

How much a sarcomere shortens depends on the load against which it is working. If contraction occurs freely, a sarcomere can shorten from 2.5 μm to 1.7 μm at maximum contraction. In a muscle cell with 20,000 sarcomeres (five cm long) this is equal to a shortening by 1.6 cm. Contraction continues until the nerve impulses cease or until the supply of ATP is used up. In contracting muscle, ATP is used up faster than it can be produced by glycolysis and mitochondrial respiration; thus, contraction cannot be maintained indefinitely.

Control of muscle contraction

Muscle contraction is turned on and off by changes in the Cat ion concentration in the sarcoplasm. Therefore, controlling the Ca^{2+} ion concentration regulates muscle contraction. A muscle contracts when a nerve impulse arrives at its sarcolemma (plasma membrane), but how does the nerve impulse cause the Ca^{2+} ion concentration to rise almost instantaneously throughout the entire sarcoplasm? Electron microscopic studies show that two membrane systems in the muscle cell, the *T system* and the *sarcoplasmic reticulum*, are the important elements.

The T System

The T system consists of tubular invaginations of the sarcolemma that penetrate deeply into the cell. These invaginations occur at regular intervals along the length of a muscle cell, usually at every Z line at multiple points around the circumference of the tubular-shaped muscle cell. When a nerve impulse reaches a muscle cell, the impulse is transferred to the sarcolernma at the point of contact between nerve cell and muscle cell, the *myoneural junction.*

From the myoneural junction the impulse is propagated over the entire sarcolemma, including the invaginations of the sarcolemma that form the transverse tubule of the T system. In this way a nerve impulse transferred to a muscle cell is almost- instantaneously propagated not only over the cell surface but along paths to the internal regions of the cell as well.

The T system ensures that an impulse arriving at one place on a muscle cell will produce virtually simultaneous responses in all the sarcomeres of all the myofibrils, causing an immediate, concerted contraction of the cell as a whole. The impulse is propagated over the sarcolemma at the same rate as a nerve impulse is conducted by a nerve axon, namely, at a rate measured in meters per second. Even though propagation down the T system is slightly slower, the result is far more rapid than would be possible if propagation depended on diffusion of molecules from the sarcolemma into the cell; diffusion would propagate a signal a few micrometers per second, a million times slower than the T system.

The Sarcoplasmic Reticulum

The sarcoplasmic reticulum resembles the endoplasmic reticulum in its structural organization. It consists of extensively branched membranous sacs that enclose spaces comparable to the cisternae of the endoplasmic reticulum. Each sarcomere contains a physically separate section of sarcoplasmic reticulum that branches to surround each myofibril at the two ends of the sarcomere. From these major portions of the sarcoplasmic reticulum thinner branches extend along all the myofibrils. At the two ends of the sarcomere the sarcoplasmic reticulum is somewhat expanded forming what are called *terminal cisternae*.

The sarcoplasmic reticulum and the T system are in close juxtaposition, usually at the Z line. The portion of the sarcoplasmic reticulum at each end of the sarcomere is in close contact with the transverse tubules of the T system. Therefore, the transverse tubules lie between the terminal expansions of the sarcoplasmic reticulum of successive sarcomeres. In a longitudinal section through a muscle cell, the end portions of sarcoplasmic reticulum and the transverse tubule often appear as three vesicles in a line, a configuration known as a triad.

The function of the sarcoplasmic reticulum is to store Ca^{2+} ions. When an impulse arriving from a nerve cell is propagated along the transverse tubule, that impulse is transmitted in turn to membranes of the sarcoplasmic reticulum, causing a great increase in Ca^{2+} ion permeability of these membranes. As a result, Ca^{2+} ions stored in the cisternae are rapidly released into the sarcoplasm, where they trigger muscle contraction. When the nerve impulses cease, the membranes of the sarcoplasmic reticulum quickly regain their impermeability to Ca^{2+} ions. The membranes of the sarcoplasmic reticulum also contain an active transport system that carries Ca^{2+} ions from the sarcoplasm into the cisternae. Therefore, the cessation of nerve impulses is rapidly followed by reaccumulation of Ca^{2+} ions in the cisternae of the sarcoplasmic reticulum, and the muscle cell relaxes.

One ATP molecule is hydrolysed by the Ca^{2+} ion transport system for each two Ca^{2+} ions pumped into the cisternae. Since muscle relaxation depends on the active removal of Ca^{2+} ions from the sarcoplasm, it is apparent that relaxation as well as contraction requires a supply of ATP.

Reconstituted contractile system and formation of structure

The main purpose of experiments in which the various muscle proteins are extracted is to identify and analyze the role of each protein through reconstitutions like those described above. However, these reconstitution experiments also provide an example of the spontaneous association of macromolecules into functional structure. Purified myosin in solution spontaneously aggregates into thick filaments under appropriate ionic conditions. These reconstituted thick filaments have the same dimensions as and closely resemble the appearance of native thick filaments in the electron microscope. Thus, not only do myosin molecules spontaneously aggregate, they do so in the correct numbers to form a thick filament of correct size. When myosin, actin, tropomyosin, and troponin are mixed in solution, they aggregate into fibers that are capable of Ca^{2+} ion-dependent contraction.

The spontaneous aggregation of the component proteins presumably also underlies the formation of structure within the muscle cell. This example, as remarkable as it is, still leaves us short of understanding fully how a complicated structure like a myofibril or sarcomere is formed during differentiation of a muscle cell.

Amoeboid Movement

Some kinds of cells move by a flowing or streaming of their cytoplasm. This form of motility is particularly striking and rapid in the many species of amoeba, and hence was long ago named *amoeboid movement*. In intact invertebrate and vertebrate animals amoeboid movement is accomplished by various kinds of cells, for example, macrophages, fibroblasts, and white blood cells. During embryogenesis many kinds of cells move in a directed fashion by amoeboid locomotion to form various tissues.

It is likely that virtually all cell types in animals are capable of amoeboid-like movement, although normally they remain stationary, at least in part because they are packed in tissues and restrained by their neighbors. The potential for amoeboid movement of tissue cells can be readily demonstrated by removing them from an animal and maintaining them in culture. All animal cells in culture, even those that fail to grow and reproduce, undergo some degree of amoeboid movement.

A major reason for the rapid progress in understanding contraction of skeletal muscle cells is that the contractile machinery is organised into well defined cytoplasmic structures, the thin and thick myofilaments, and the like. The structures serve as a guide for analysis of function. Smooth muscle cells also contain large quantities of much the same contractile machinery as do striated muscle, but the proteins for contraction are not so clearly organised into cytological structures. Consequently, it has been far more difficult to analyse the molecular basis of contraction of smooth muscle cells.

Similarly, amoeboid movement has been difficult to study because it is not based on a well defined cytological structure. However, the detailed information we now possess about the molecules involved in muscle contraction has, during the last several years, formed the basis for a molecular explanation of amoeboid movement. Amoebae are capable of a surprisingly rapid traverse across a surface. Larger species can move several hundred μm per minute.

The cytoplasm of an amoeba is organised into an outer *ectoplasm* that is a gel. It contains networks of actin microfilaments that are likely responsible for the gelled state. The gelled ectoplasm surrounds the fluid portion or sol part of the cytoplasm, called the *endoplasm*, in

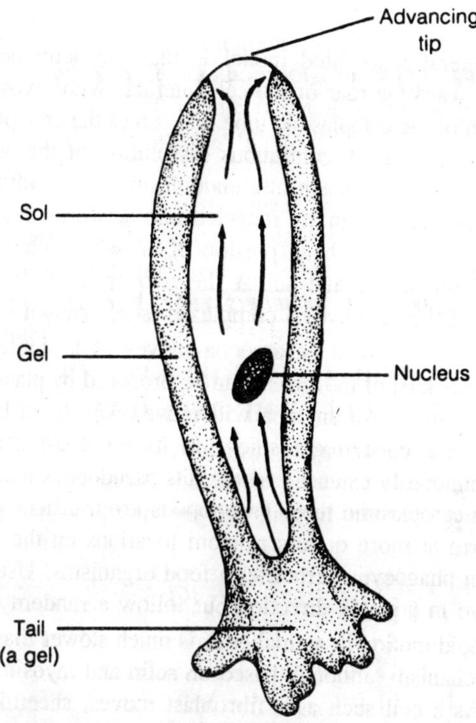

Figure 9.10 : Diagram of the gel-sol structure of an amoeba. Cytoplasm in the sol state flows forward enclosed in a tube of cytoplasm in a gel state. At the advancing tip cytoplasm changes from sol to gel, and in the tail region cytoplasm is converted from gel to sol. The nucleus tumbles along with the stream of cytoplasm, but remain; at the same relative position a little behind the middle of the cell.

which the nucleus, mitochondria, food vacuoles, and other cytoplasmic organelles tumble along during amoeboid streaming. The ectoplasm forms a rigid tube through which the endoplasm flows. As the endoplasm flows in the tube, it is converted into a gelled ectoplasm at the forward tip of the amoeba, thereby continuously extending the ectoplasmic tube. At the back end of the amoeba, gelled ectoplasm is converted into sol that then flows forward in the endoplasm.

Where is the motive force for amoeboid movement generated? Looking at a moving amoeba immediately gives the impression that the motive force for the for ward streaming of the endoplasm is produced by steady contraction of the ectoplasm near the back end of the cell. This is equivalent to an amoeba squeezing itself forward. The contractile mechanism that propel the endoplasm forward is believed to consist of actii and myosin.

The generally accepted model is that the actin network in the ectoplasm toward the rear of the cell interacts with myosin to produce a contraction of the ectoplasmic tube that forces the endoplasm forward. This is accompanied by continuous dissolution of the actin networks and conversion of ectoplasm into endoplasm. At the advancing end of the cell, actin and myosin are recycled into networks that later interact in the ectoplasm in the back portion of amoeba. What regulates the interconversion of gel and sol at the two ends of the cell and the contraction of the actomyosin complexes is not known.

In its simplest form an amoeba moves as a single tube as just described. This kind of movement can be produced by placing an amoeba in a beam of light. An amoeba will move away from light (*negative phototaxis*) in a concerted fashion. In its more usual movement an amoeba continuously extends and retracts pseudopods that are branches of the main ectoplasmic tube. Pseudopods are transient structures that seem to form at more or less random locations on the cell, but they also form in phagocytic response to food organisms. Usually amoebae do not move in a single direction but follow a random walk pattern.

Amoeboid motion in animal cells is much slower than in amoebae, and the mechanism, although based on actin and myosin, is somewhat different. As a cell such as a fibroblast moves, sheetlike projections called *lamellipodia* are formed at its leading edge. The phenomenon is called *ruffling*. As the cell moves forward, the lamellipodia are displaced backward from the leading edge of the cell and disappear a short distance behind the leading edge.

Lamellipodia are gel structures rich in actin networks. Their formation and disappearance reflect gel-sol interconversion similar to that seen in amoebae. It is believed that the actin fiber networks in lamellipodia also contain myosin, and the interaction of myosin with actin networks creates a contractile force that pulls the cell forward in a manner that is still poorly understood.

CYTOPLASMIC STREAMING IN PLANT AND ALGAL CELLS

Plant and algal cells and other eukaryotic cells with rigid cell walls are incapable of amoeboid movement because they are physically restrained. However, the cytoplasm in these kinds of cells generally does move, a process called *cyclosis*. In some plants cyclosis is rapid, carrying cytoplasmic particles at rates up to 100 μm per second. Cytoplasmic streaming has never been observed in prokaryotes.

Giant, multinucleated algal cells such as *Nitella* and *Chara* are

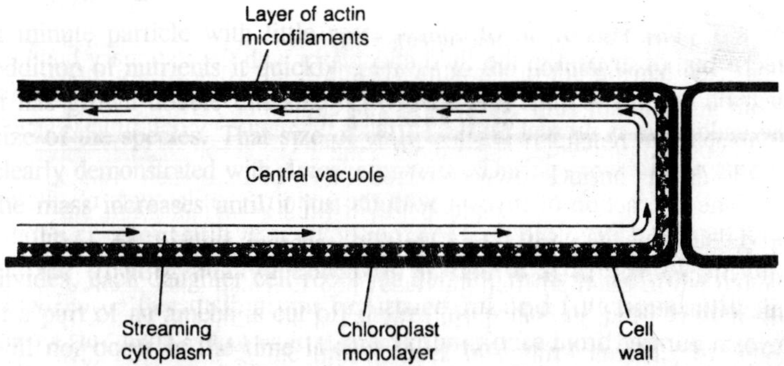

Figure 9.11: Drawing of a cell of the alga Nitella. Chloroplasts are arranged in a monolayer embedded in a cytoplasmic gel. A layer of actin filaments forms an interface between the chloroplasts and the streaming cytoplasm.

especially favourable for the study of cytoplasmic streaming. The cytoplasm immediately adjacent to the inner side of the plasma membrane is organised into a thin cortex or ectoplasmic gel. The cortex contains large numbers of chloroplasts immobilised as a single layer. Most of the internal part of the cell is occupied by a giant central vacuole. Between the central vacuole and the cortex is a fluid endoplasmic region that undergoes continuous, unidirectional streaming at rates up to 100 μm per second.

Mitochondria, nuclei, vesicles of various sizes, and other cytoplasmic organelles, are carried along in the stream. Streaming follows a fixed curved path along the length of the cylindrical cell, across the end of the cell, and back along the opposite side of the cell in the opposite direction, then around the other end, forming an endless moving belt. Between the oppositely moving streams is a narrow region of stationary, fluid endoplasm.

The route of streaming is determined by a layer of bundles of actin microfilaments bound to the inner surface of the cortex. These bundles are stationary and are arranged in parallel with the path of streaming. Each bundle contains 50 to 100 microfilaments.

All the actin microfilaments are arranged with the same polarity (as in thin myofilaments of striated muscle), and this polarity determines the direction of streaming. The arrowheads produced by binding of myosin fragments point upstream. When the layer of microfilament bundles is experimentally disrupted in a localised region, for example, by microbeam irradiation, streaming stops or becomes passive in that region.

The force behind cytoplasmic streaming in *Nitella is* produced by interaction of myosin in the fluid endoplasm with the stationary bundles of actin microfilaments on the inner surface of the cortex. Myosin in the endoplasm is present on the surfaces of minute cytoplasmic vesicles of various sizes.

The interaction of myosin with the actin microfilaments results in propulsion of the vesicles, apparently by a walking motion along microfilaments in a direction governed by the polarity of the microfilaments. When plastic beads (0.7 μm in diameter) coated with myosin purified from rabbit muscle are applied to the inner surface of the ectoplasm exposed by breaking open a cell, they move along the surface of the ectoplasm in a direction predicted from cytoplasmic streaming in the intact cell. Movement requires the presence of ATP. The propulsion of vesicles along the inner ectoplasmic surface is believed to cause the streaming motion of the entire endoplasm.

MOVEMENT BY CILIA AND FLAGELLA

Muscle contraction, amoeboid movement, cytoplasmic streaming, and cytokinesis are all achieved by interaction of myosin with actin. The movement of cilia and flagella is based on an entirely different pair of macromolecules, tubulin and dynein. However, the principle behind the movement is essentially the same; one kind of molecule, myosin or dynein, moves along a filament composed of the other macromolecule, that is, movement of myosin along an actin microfilament and movement of dynein along a microtubule.

Cilia and eukaryotic flagella have basically the same structure, and these two organelles represent two versions of a single mechanism for motility. Cilia are present in large numbers on cell surfaces, whereas cells that move by means of flagella usually possess a single flagellum or a pair of flagella. Flagella are generally much longer than cilia. Cilia range from 0.2 μm to ten μm in length, and flagella range from ten μm to several mm. Both have a diameter of 0.2 μm. The flagella of eukaryotes are fundamentally different in both structure and mode of operation from the flagella of prokaryotes. Bacterial flagella are discussed in the final section of this chapter.

Cilia occur in a wide range of cell types among the protozoans, invertebrates, and vertebrates. They occur more rarely among algae, plant cells, and fungi. One of the major groups of protozoa, the *Ciliata*, consisting of many thousands of species, is partially defined and named by the possession of ciliary motility. Familiar examples are *Paramecium* and *Tetrahymena*. The cilia of most ciliated protozoa provide the means of cell motility but, of equal importance, are used to create currents to carry food organisms such as algae, bacteria, and other protozoa into an oral cavity where they are ingested by phagocytosis. Ciliated epithelial cells are common in animals, where in most cases they serve to propel fluid across the surface of epithelial tissues. For example, portions of the respiratory tract of mammals are lined with a ciliated epithelium whose constant ciliary activity clears mucus, dust, bacteria, and so forth, from the respiratory passages.

Flagella are also widely distributed among eukaryotes. A major group of protozoa, the *Flagellata*, is defined by the presence of flagella. Many kinds of unicellular algae, for example, *Euglena* and *Chlamydomonas*, possess flagella. The sperm cells of most invertebrate and vertebrate animals and of a few plants are also propelled by flagella.

Structure of Cilia and Flagella

Cilia and flagella are sufficiently alike that one description of structure can serve for both organelles. Cilia are remarkably constant in structural composition in eukaryotic species; the cilia of *Paramecium* are essentially identical to the cilia in the feeding apparatus of a clam and in the ciliated epithelium of the oviduct in a mammal. Microtubules form the core of cilia and flagella and run the length of these organelles. Nine pairs of microtubules are arranged in cylindrical fashion around two central microtubules. This 9+2 pattern is readily seen by electron microscopy in a cross section through the cilium. The set of microtubules, called collectively the *axoneme (axo* = axis; *neme* = threads), is enveloped in a cylindrical outpocketing of the plasma membrane. Hence, a cilium is a fingerlike, microtubule-containing projection of the cytoplasm.

At the base of a cilium is a structure called the *basal body* or *kinetosome*. The basal body has the same structure as a centriole, and these two organelles are considered to be one and the same. Indeed, the centriole in a developing sperm cell moves to the cell surface, where its component microtubules act as initiation sites for polymerization of the longer microtubules that form the flagellum, that is, the centriole becomes the basal body. The centriole is normally associated with the mitotic spindle organising center (centrosome) in animal cells. The basal body of a cilium or flagellum, like a centriole, is made up of 9 triplets of microtubules arranged in a circle to form a cylinder about 0.15 μm in diameter and about 0.5 μm long. In contrast to a cilium, there are no microtubules in the center of a centriole. The basal body is believed to serve as a starting or nucleating organelle in the formation of a cilium or flagellum. In the mature sperm cells of some insects the basal body is absent, suggesting that once a flagellum has been formed, the basal body is not needed for flagellar function.

MOLECULAR MECHANISM OF CILIARY AND FLAGELLAR MOVEMENT

A variety of experiments has provided an explanation for the beating of cilia and flagella, although important aspects of the mechanism remain to be clarified. In an early experiment cilia were observed to continue to beat and independently move through the medium for a short time after being severed from the surface of a ciliated protozoan. Beating ceased when an isolated cilium had used up its energy reserves.

The experiment proved that ciliary movement is not achieved by a motor at the base that propels the cilium back and forth; rather, the cilium itself contains the mechanism for movement. Treatment of isolated cilia with a mild detergent removes the plasma membrane, leaving the naked *axoneme* with the 9+2 pattern of microtubules. When the proper inorganic ions and ATP are added to the solution, the naked axoneme beats. Thus, it is clear that the plasma membrane around the cilium is also not necessary for movement, and second, the source of energy for movement is ATP, as in systems of motility based on a myosin-actin interaction.

A similar experiment can be done with flagella removed from sperm. As long as ATP is supplied in the medium, the flagella continue their wavelike beating. In ciliated cells such as *Tetrahymena*, mitochondria are concentrated just under the cell surface near the bases of cilia, a location favourable for provision of ATP to the cilia. In the sperm of most animals and plants the first part of the axoneme is surrounded by a single large mitochondrion or a group of mitochondria, presumably to supply ATP for flagellar movement.

The ATPase that hydrolyzes ATP to ADP + P_i to provide energy for ciliary beating can be extracted from axonemes and still leave the axonemes intact. This ATPase has been named *dynein*, and its location in the axoneme was discovered in the following way. In normal axonemes a pair of projections or arms occurs at regular intervals on one side of each doublet of microtubules in the axoneme. In axonemes from which dynein has been extracted the arms are missing. If dynein is allowed to reattach to the axonemes, the arms and ATPase are simultaneously regained by the axonemes.

In a rare, genetically inherited disease described in human males

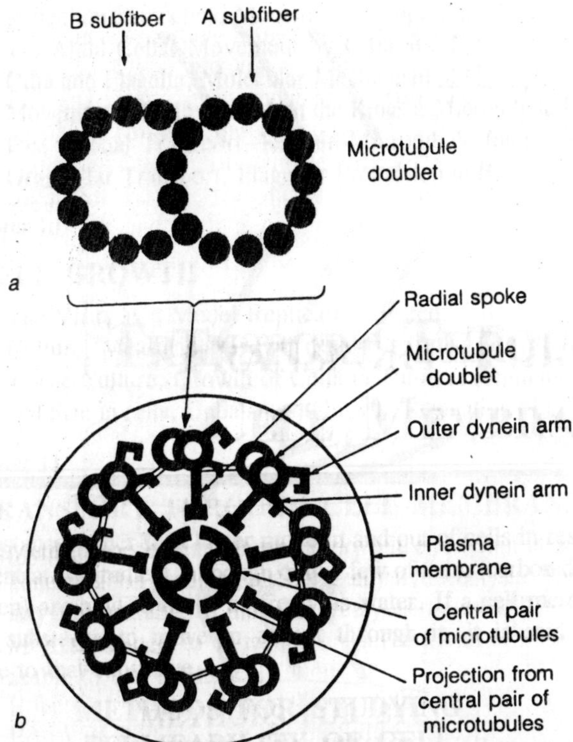

Figure 9.12: (a) A microtubule doublet consisting of an A subfiber with 13 protofilaments of tubulin and a B subfiber with 11 protofilaments. (b) A cilium contains nine microtubule doublets and a central pair of microtubules. The dynein arms provide for sliding of one doublet along another during the beating of the cilium. The functions of radial spokes and the projections from the central pair of microtubules have not been as well defined.

the dynein arms are abnormally short or totally absent. A mutation is present in a gene carried by the X chromosome that causes synthesis of abnormal dynein or that causes a failure of dynein to attach normally to the microtubules.

The cilia and flagella in these individuals are incapable of movement. As a result, the spermatozoa, whose flagella are normal in all structural aspects except for the lack of dynein, are nonmotile, and the individuals are infertile. In addition, the cilia of the epithelial cells lining the respiratory tract, paranasal sinus, and the Eustachian (auditory) tube are nonmotile. Therefore, 'these persons do not possess the normal mechanism for clearing dust, and debris, including viruses and bacteria, from the affected area, and they chronically suffer from bronchitis, sinusitis, common colds, and ear infections.

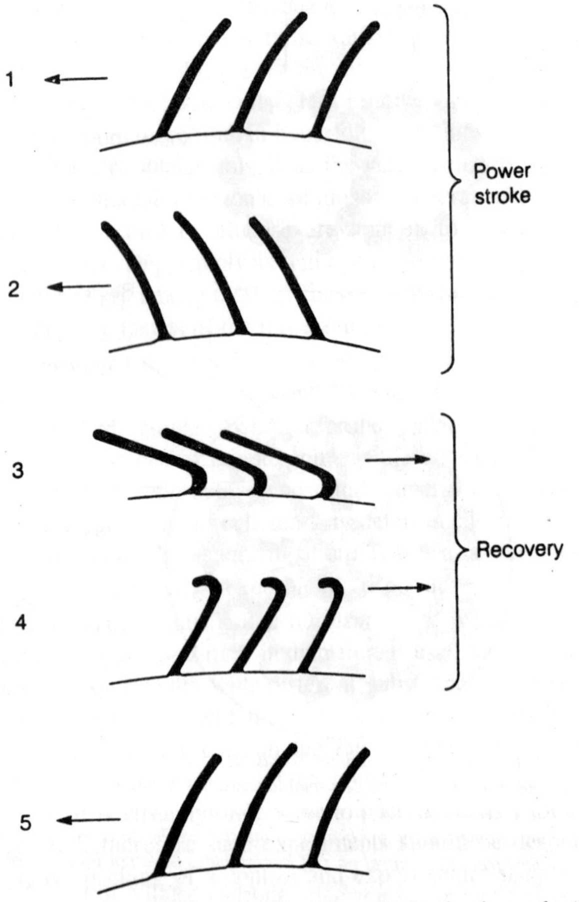

Figure 9.13: Cycle of ciliary stroke shown for three cilia beating in synchrony. (1) Beginning of the power stroke. The cilia are straight and stiff. (2) Half completion of the power stroke. (3) Start of recovery stroke. The cilia are flexible, and by bending reduce frictional drag. (4) Near completion of the recovery stroke. (5) Start of the next power stroke.

Microtubules in general are constructed of 13 protofilaments made of tubulin. However, because the pair of microtubules in each of the nine doublets of the axoneme is partially fused in a lengthwise direction, the number of protofilaments is different. One of the two microtubules, called the A *subfiber*, contains a full set of 13 protofilaments. The *B subfiber* has only 10 or 11 protofilaments. At the site where the two or three protofilaments are missing, the B subfiber fuses with the A subfiber. Dynein molecules form pairs of arms on the A subfiber. The arms form transient bridges to the B subfiber in the next doublet

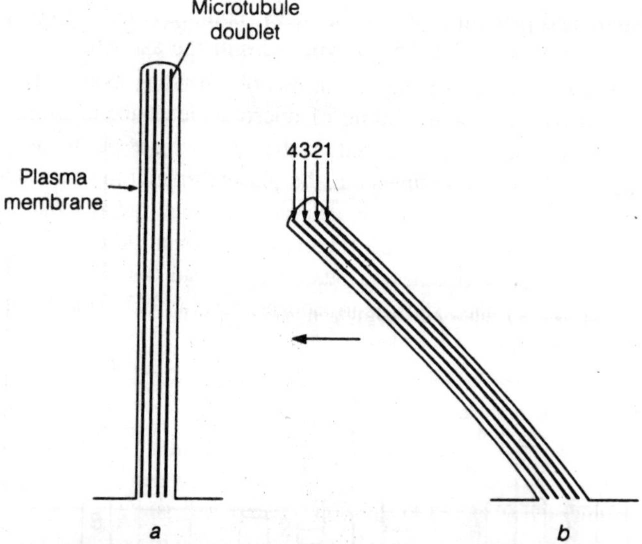

Figure 9.14: Sliding of one microtubule doublet along another causes a cilium to bend. (a) The cilium is perpendicular to the cell surface; all microtubule doublets extend the full length of the cilium. (b) The cilium is bent by a slight active sliding of doublet number 1 baseward along doublet number 2, number 2 baseward along number 3, and number 3 baseward along number 4. The doublets all remain the same length. However, they end at different points relative to the tip of the cilium because of the slight displacement of one doublet against another.

during beating of the cilium. Each of the nine doublet microtubules is connected by radial spokes to the central pair of microtubules. Spokes are present at regular intervals along the length of the axoneme, and curved rods project from the central pair of microtubules at regular intervals.

Over 250 proteins have been identified by electrophoresis from purified axonemes that have been dissociated with a strong ionic detergent. Two of these proteins, α- and β-tubulin, account for about 70 percent of the total protein in an axoneme. The two dynein arms are each composed of multiple polypeptides. The outer arm contains about nine different polypeptides, two of which are ATPases that split ATP to ADP + P_i during generation of a ciliary beat.

The inner arm is structurally similar, but seems to be made from completely different polypeptides. Each radial spoke contains about 17 different polypeptides. The majority of axonemal polypeptides are unaccounted for. A few may be microtubule- associated proteins (MAPS), and many are probably present in the matrix spaces between the recognizable structures. Except for the ATPases of dynein arms and for α- and β-tubulin of microtubules, nothing is known about the function of the more than 250 polypeptides of the axoneme.

The power stroke of a cilium consists of a bending motion. Bending is achieved by lengthwise sliding of microtubule doublets along each other. Sliding has been observed in the following experiment. Cilia were detached from *Tetrahymena,* the plasma membrane surrounding them was removed with nonionic detergent, and the radial spokes were broken by brief treatment with the enzyme trypsin. Addition of ATP did not cause the axonemes to beat, but instead doublets slid along each other, causing the axoneme to elongate several fold. This elongation happens as follows.

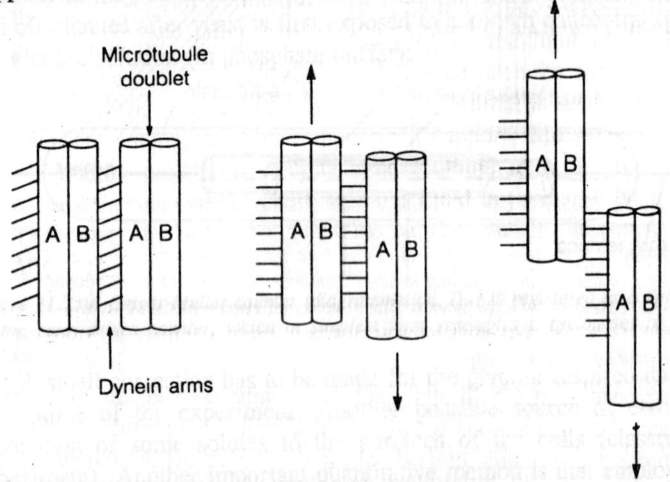

Figure 9.15: Diagram of the response of microtubule doublets in an isolated axoneme. One doublet continues to slide along another, causing the axoneme to elongate several fold. Eventually the doublets slide completely off one another.

The dynein arms are permanently attached to the A rnicrotubule in each doublet. At rest the arms are tilted downward toward the base of the cilium, as they form bridges to the B microtubule of the next doublet. The addition of ATP causes the dynein arms to move to a 90° angle relative to the A microtubule, which causes the two doublets to slide relative to each other. At the end of the stroke dynein detaches from the B microtubule, bends downward, reattaches to the B microtubule and repeats the stroke. In isolated axonemes repetition of the cycle eventually results in sliding of one doublet off the end of the other. In a cell the microtubule doublets are anchored at the base of the cilium. Hence, slight sliding of one doublet along another causes them to bend.

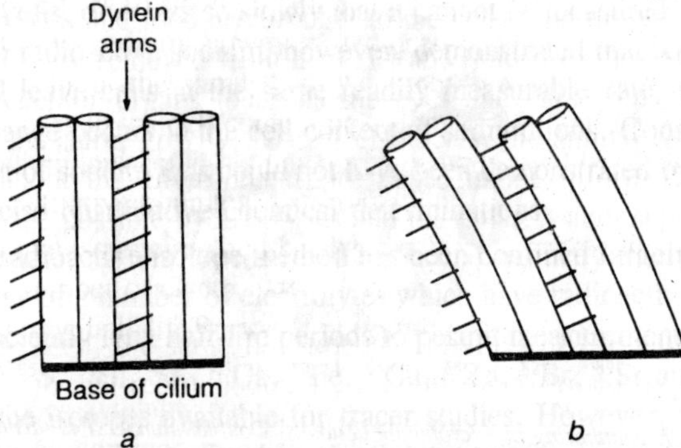

Figure 9.16: Sliding of one microtubule along another in an intact cilium causes the cilium to bend, creating a power stroke.

Sliding mediated by dynein arms is the mechanism for generation of forces within axonemes, but how are these forces organised to produce the ciliary beat? The nine pairs of doublets are numbered 1 through 9 by their orientation relative to the central pair of microtubules. To define this orientation a line is drawn through the centers of the central microtubules. A second line drawn between the central microtubules at a 90° angle to the first line bisects one of the outer doublets. This outer doublet is number 1; the dynein arms of doublet 1 point toward doublet 2, the arms of 2 point toward doublet 3, and so on, around to doublet 9.

All the dynein arms on all the doublets generate force in the same direction, and therefore every doublet slides in the same way. If all the dynein arms on all of the doublets were to operate at once, the resultant forces would almost cancel, resulting only in a twist of each doublet. This may be difficult to visualize, but note that doublet 1 would have two equal and almost opposite forces applied to it: doublet 2 would force it toward the base of the cilium and doublet 9 would force it toward the tip.

This dilemma appears to be solved by dividing the doublets into two functional groups. Doublets 1 through 4 are thought to act together during the power stroke in forward swimming, causing a force that tends to elongate the cilium on one side. The elongation is accommodated by bending of the cilium. Once the power stroke is done, doublets 5 through 9 slide against each other to straighten the cilium. In some way the operation of the two groups of doublets must be different since the cilium is rigid in the power stroke but bends in the recovery stroke. How the two groups of doublets are controlled to achieve the proper temporal pattern of sliding is not known. The two groups of doublets

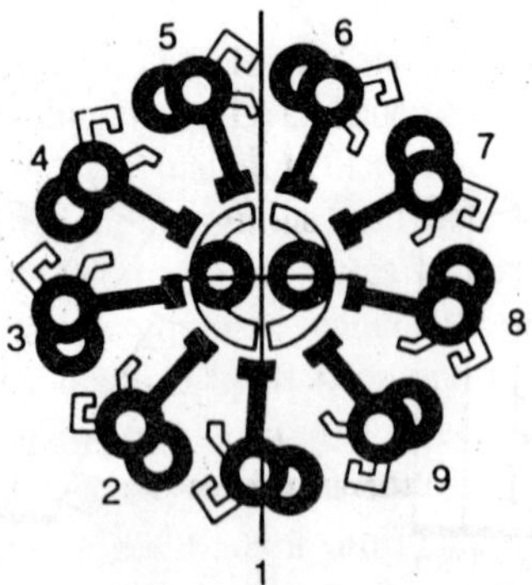

Figure 9.17: Numbering of microtubule doublets. Doublet number 1 is identified by a line that is perpendicular to a line that passes through the centers of the central pair of microtubules. Numbering 2 to 9 is in the direction of the dynein arms.

must operate in the opposite order when the direction of the power stroke is reversed in backward swimming of the cell.

Finally, to accomplish cell movement of a motile cell or movement of fluid across the surface of a stationary cell, the cilia must all beat in approximate synchrony, perhaps by signals transmitted by the plasma membrane or by the complex network of fibers that interconnect all the cilia at their basal ends just inside the cell surface.

MOTILITY BASED ON THE KINESIN-MICROTUBULE INTERACTION

Vesicles, pigment granules, and other particulate structures in the cytoplasm often move rapidly from one part of the cell to another by a mechanism that does not involve cytoplasmic streaming. A dramatic example occurs in the rapid change in colour of some species of fish. Fish colouration is due to large specialised cells called *chromatophores* that are present in the scales.

In response to stimulation some fishes can switch from a pale appearance to bright or dark colouration in a few seconds. When the pigment granules it the chromatophores are all tightly gathered into a knot in the center of the cell, the fish has very little colouration. On stimulation, the granules rapidly disperse throughout the cell along a radially arranged network of microtubules, anc the cell and the fish become côloured. The movement of pigment granules is rapid, directional (along microtu· bules), and reversible.

Another example of rapid intracellular movemen is the migration of vesicles along axons of nerve cells Materials needed to maintain the structure and funo tion of the axon of a nerve cell are synthesized in the body of the cell and must be transported down the axon. Substances that are released from the end of the axon in transmission of a nerve impulse to anothei nerve cell at a synapse or at a muscle. nerve junction are carried the full length of the axon Some axons are more than two meters long.

Fast Axonal Transport

The squid contains an axon of particularly large diameter, which is favourable for observing axonal transport. Small vesicles move away from the cell body (*anterograde movement*) and back toward the cell body (*retrograde movement*) along microtubules. Anterograde movement delivers materials down the axon, and retrograde movement probably represents return of emptied vesicles for recycling.

Vesicle movement is rapid-several μm per second-and is called *fast axonal transport*. Fast directional transport of vesicles along microtubules occurs in many kinds of cells other than axons, for example, fibroblasts. Organelles, including particles other than small vesicles, are transported. The mechanism of transport is likely the same in these various situations, and the name *fast organellar transport* is more appropriate than the term *fast axonal transport*.

Kinesin-Microtubule Interaction in Fast Organellar Transport

Current understanding of the molecular mechanism of fast organellar transport has come largely from study of the giant axon of the squid. The cytoplasm obtained from axons (axoplasm) has been separated into soluble supernatant and organelle-enriched fractions by centrifugation. Addition of the organelles to purified microtubules does not result in organelle movement, but when the soluble fraction from squid axons is added along with ATP, the organelles adhered to and migrated along individual microtubules.

A translocator protein has been purified from axoplasm and from bovine brain and given the name kinesin (from the Greek word *kinein,* to move). Latex beads (0.15 μm diameter) coated with purified kinesin are observed to move along microtubules in the presence of ATP. Kinesin, like myosin and dynein, is composed of a high molecular weight complex of polypeptides. Like dynein, it binds to microtubules (myosin binds to actin microfilaments), but it differs from dynein in several major ways and represents a different (a third) mechanochemical mechanism of movement.

The movement generated by kinesin along microtubules is unidirectional, as in the case of myosincoate 1 beads along actin microfilament bundles. Unidirectional movement probably reflects the polar nature of microtubules. However, in reconstitution experiments in which the soluble fraction of axoplasm is mixed with microtubules, vesicles have been observed to move in opposite directions along a single microtubule. Since purified kinesin produces only unidirectional movement, this suggests that yet another translocator protein may be present that generates movement in the opposite direction.

In axons and in the cytoplasm of other cells, larger vesicles and organelles (>0.25 μm in diameter) undergo *saltatory*, or jumping motions; vesicles and organelles move rapidly, stop transiently, and then move again, giving the impression of a jumping action. Smaller vesicles, for example less than 0.1 μm in diameter, move continuously without jumping. In isolated axoplasm the large vesicles and organelles move continuously and uninterruptedly along microtubules in the same manner and speed as the smaller ones.

It may be that the saltatory movement is caused by cytoskeletal structures that transiently get in the way of larger vesicles and organelles but do not impede smaller ones. Alternatively, saltatory movement may be the result of periodic dissociation of particles and vesicles from microtubules. Dissociation is consistent with the observation that the rate of translocation of a vesicle during a single jump is the same as the rate of continuous nonsaltatory movement of smaller vesicles.

Pigment granules may move by means of kinesin or a kinesin-like protein, since movement is along microtubules and is rapid. Kinesin has also been identified in the mitotic spindle and may be responsible for the chromosome movement in anaphase.

FLAGELLAR PROPULSION IN BACTERIA

Many species of bacteria move by means of flagella. Others move by gliding across a surface, but little is known about the mechanism of gliding, and it will not be discussed. Some bacteria have a single flagellum, and others have several; and yet other bacterial species have dozens of flagella that project over the entire surface of the cell.

In *E. coli* at least 35 genes are involved in the assembly, structure, and function of a flagellum. One or more genes involved in flagellar formation in *E. coli* is subject to catabolite repression. In the presence of a plentiful supply of glucose, the synthesis of flagellar components is repressed. This underscores the importance of flagellar motility in cell nutrition. Recall that *E. coli* is a heterotroph that is able to synthesize all of its organic components from glucose. In the presence of glucose the need for motility is obviated, and flagella are not formed as cell reproduction proceeds.

Bacterial flagella are fundamentally different from eukaryotic cell flagella in both structure and function. A bacterial flagellum is about 20 nm in diameter, or about 1/20 the diameter of a eukaryotic cell flagellum, and is usually 10 to 20 μm long. Because of its small diameter, special optical methods must be used to see the beating flagellum of a living cell. The flagellum is not enclosed in an outpocketing of the plasma membrane, but instead projects from the cell as a naked protein filament. It is anchored in the plasma membrane by means of a base complex.

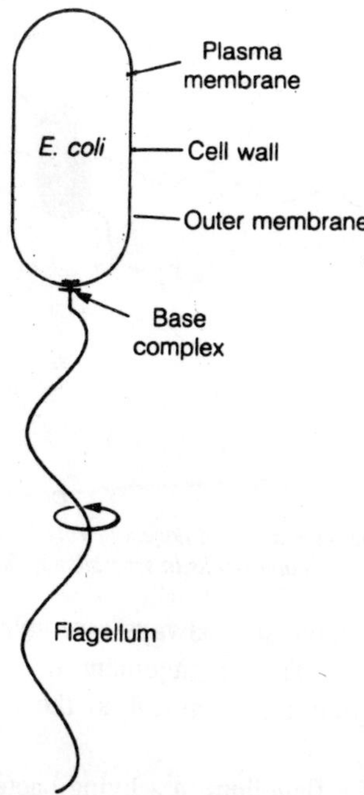

Figure 9.18: The flagella of E. coli are anchored in the surface of the cell by a base complex. Only one of the six flagella is shown.

When multiple flagella are present, they form a bundle and work cooperatively to propel the cell forward in a smooth swimming motion. Reversal in the direction of rotation of the flagella results in separation of the flagella from the bundle, causing a tumbling motion of the bacterium.

The bacterial flagellum is constructed of a protein known as *flagellin* with a molecular weight of about 53,000. The flagellum is made of five long chains of flagellin molecules wound around one another in a tight helical arrangement (a *pentahelix*), creating a tubular structure. Pure flagella can be prepared by severing them from bacteria. When purified flagella are gently heated or exposed to low pH, they dissolve, yielding a solution of flagellin. If the dissociating conditions are reversed, the flagellin molecules reaggregate into flagella. For this to happen, the solution of flagellin must be seeded by adding short fragments of flagella, which can be obtained by breaking up intact flagella with sound waves.

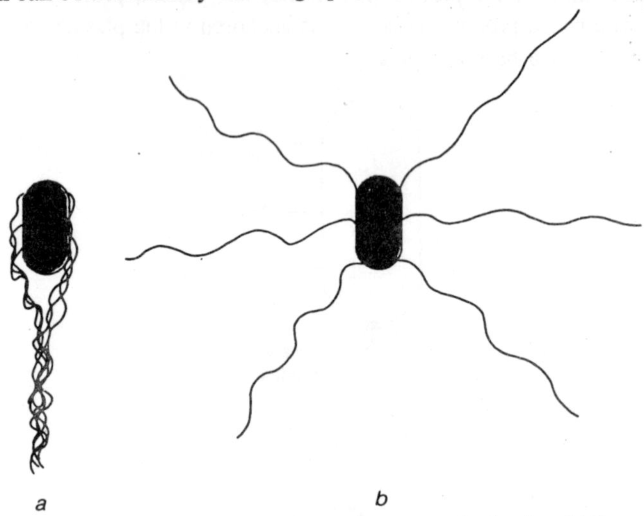

Figure 9.19: (a) Cooperative action of flagella during forward swimming. (b) Reversal in the direction of flagellar rotation results in separation of flagella and causes the bacterium to tumble.

The flagellin molecules always assemble at only one end of the flagellar fragment. The arrangement of flagellin molecules in a reconstituted flagellum is identical to the arrangement in a natural flagellum.

Formation of a flagellum in a living bacterium occurs by addition of flagellin molecules to the tip of the flagellum. The individual flagellin molecules pass down a channel in the center of the pentahelix that forms the flagellum and are assembled onto the distal ends of the five chains. The cilia and flagella of eukaryotes also grow by addition of protein molecules that pass down the organelle and assemble at the tip.

The bacterial flagellum is rigidly held in an inflexible helical form that is much like a corkscrew. Mutant bacteria have been obtained in which the flagellum is a straight, rigid rod. These mutants are immotile since the pushing or pulling action of the flagellum depends on a helical

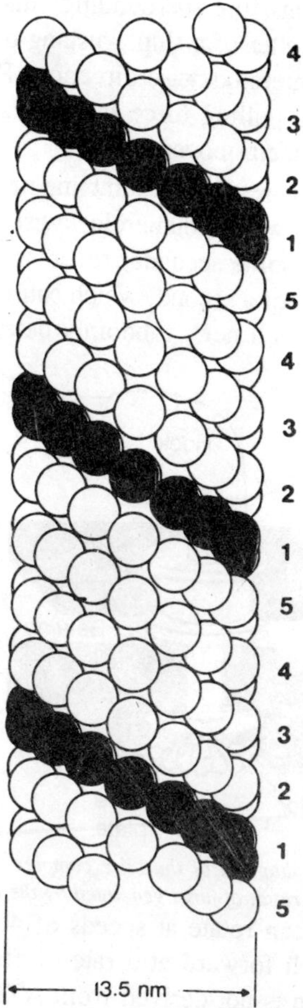

Figure 9.20: Five chains of flagellin molecules coil around a central axis to form a bacterial flagellum.

shape. The mutation is undoubtedly in the gene for the flagellin protein and causes a shift in the interaction of flagellin molecules with one another, so they form a straight structure rather than a helical one.

Propulsion of a bacterium occurs when the helical flagellum is rotated at the point of insertion into the cell. The force creating rotation is applied in the cell to the base plate (in the plasma membrane) to which the flagellum is attached. This is a totally different arrangement from that in the flagella and cilia of eukaryotes, where force is generated along the entire length of the structure.

Rotation of the rigid, helically twisted bacterial flagellum exerts a pushing force against the surrounding fluid, which drives the cell forward like the propeller of a ship. Pushing occurs when the flagellum is rotated in the counterclockwise direction. Rotation in the clockwise direction results in a pulling force by the flagellum, although this is not efficient in producing movement.

Clockwise rotation results in a tumbling motion of the cell. Normally a bacterium swims forward (counterclockwise rotation of the flagellum) in a smooth line for some seconds, reverses the direction of flagellar rotation for a fraction of a second, which causes a tumbling motion, and then swims forward in a new, randomly determined direction.

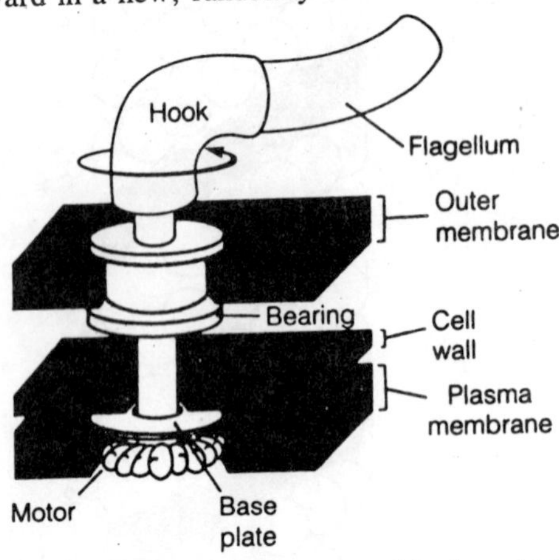

Figure 9.21: Schematic diagram of the base complex of the bacterial flagellum. The flagellum is rotated as a result of force generated by the motor in the plasma membrane.

The flagellum can rotate at speeds of 40 revolutions per second, which drives the cell forward at a rate of 50 μm per second. Energy for flagellar rotation is not derived from ATP, but rather comes from the proton gradient across the plasma membrane, generated by oxidative phosphorylation. Remember that in the bacterial cell the machinery for oxidative phosphorylation is in the plasma membrane, which is also the location of the flagellar motor.

The overall migration of motile bacteria is not necessarily random in direction. For example, photosynthetic bacteria move in the direction of greater light intensity, a behaviour known as *positive phototaxis*. At least some motile species have receptors in the plasma membrane for nutrients such as amino acids and sugars that allow them to move toward higher concentrations of these nutrients, a process called *positive chemotaxis*. The bacteria are not able to sense concentration gradients directly but move toward higher concentrations of nutrients by a trial-

and-error method. When environmental conditions are constant, the bacterium swims in a straight path for a time, stops, tumbles, and swims another straight course in some other direction. When a bacterium encounters a gradient of an amino acid, the binding of the amino acid to receptors in the plasma membrane causes the cell to swim a straight course for a longer period before tumbling.

If the course of swimming has taken the bacterium to a still higher concentration of the amino acid, its receptors signal it to swim on the straight course for longer than usual once again. However, if the bacterium has chanced to move to a lower concentration, fewer receptors are occupied, and the cell swims forward for a shorter course. Monitoring the concentration of the nutrient over time and swimming a longer straight path when the concentration of attractant is increasing results in net movement of the bacterium up a concentration gradient.

By a similar process of monitoring concentration over time, bacteria are able to migrate away from unfavourable conditions such as toxic concentrations of acetate, ethyl alcohol, and low or high pH. The process of migrating away from unfavourable conditions is called *negative chemotaxis*.

10

CELL GROWTH

The cell grows from within (by intussusception). Growth occurs only if temperature, pH, oxygen and other conditions are favourable, and if the required nutrients and adequate water are available. In the absence of water, the cell may pass into such dormant resting stages as cysts or spores, and remain thus for many years. If nutrient is lacking, or if certain required nutrients are limiting factors, growth may stop or become abnormal. Growth is also easily inhibited by poisons, radiations and, in fact, by all agents that injure the cell or interfere with normal cell functions.

THE VIRUS AS A MODEL REPLICATING SYSTEM

Under favourable conditions a virus particle will have produced between 100 and 300 replicas of itself in half an hour after entering a host cell, each replica consisting of a DNA molecule (chromosome) and a covering of protein. As soon as the virus enters the host cell it usually represses the synthesis of host nucleic acids, starts replication of its own DNA and directs transcription of its genes to form messenger RNA.

The proteins synthesized under direction of the virus mRNA are synthesized on the host ribosomes, using the soluble RNA of the host cell. Among the proteins synthesized by the virus are the packaging proteins for the viral coating and enzymes such as the RNA polymerase necessary for mRNA synthesis and the enzymes facilitating entry of the virus into the host cell. The substances repressing the host DNA synthesis may be proteins also. Although the virus has given much insight into our knowledge of DNA and RNA structure and function which has helped in the solution of the genetic code, it offers less help in understanding many of the problems of cell growth.

LIFE CYCLE OF A CELL CULTURE

One bold viewpoint holds that unicellular organisms are potentially immortal and that in multicellular organisms only cells of the "germ plasm" which are passed during reproduction, from generation to generation, are potentially immortal, but the specialised cells of the body (soma) must die.

It has been thought in the past that a stock of protozoans always grows old and dies after prolonged subculture. Thus individuals of the protozoan *Paramecium bursaria,* for example, isolated from cultures which had been kept in the laboratory for many years, divide more and more slowly with time, often contain nuclei in the process of degenerative fragmentation and ultimately die out. Similar life cycles had been observed in other species of protozoa, and attempts were made to determine the span between youth, old age and death for various kinds of protozoans.

It was also demonstrated that, after conjugation, some individuals in each of many kinds of protozoans were rejuvenated and more viable; presumably as a result of a favourable genotype. It was then concluded that unicellular organisms age and ultimately die unless periodically rejuvenated by some type of nuclear reorganisation. But it was soon realised that this conclusion is not valid, because it is based on observations of protozoans grown in a medium which was not controlled.

To control conditions of growth, it is necessary to grow the protozoans in the absence of any other microorganisms (axenic culture) and preferably on a synthetic nutrient known to be fully adequate. *Tetrahymena pyriformis (geleii)* grown under such conditions has been kept in axenic culture for years by Lwoff, Kidder and Dewey, growing at a rate even in excess of the wild state and without loss of vigor.

It is of interest that conjugation has never been observed in any of these cultures, yet wild stocks of *Tetrahymena,* isolated from ponds, undergo conjugation each time the proper mating types are mixed. Similar experiments with a variety of other kinds of microorganisms in repeated transfer or in steady state growth devices show that many microorganisms may be grown at a constant growth rate for thousands of generations. To that extent it may be stated that microorganisms are probably potentially immortal and that growth of protoplasm and cell division follow without abatement if nutrient is available and other conditions are favourable. Yet a growth cycle ending in death will occur in a pure culture of microorganisms if they are grown in a limited or unbalanced environment, or under other unfavourable conditions.

For example, if a single vigorous protozoan is inoculated into a limited volume of a known culture medium, it divides regularly until lack of food limits its growth. A plot of the number of divisions against time gives a characteristic curve. After the initial rise the curve flattens out and forms a plateau (the stationary phase) as food becomes a limiting factor.

The curve declines as the individuals begin to die because of inanition. If an identical medium is initially inoculated with a starving individual, a lag precedes the rising straight line, representing the period of recovery from the effects of previous starvation. If individuals are removed from a culture in which they are about to die for lack of food, and are inoculated into fresh medium identical to the first one, after a lag phase they resume growth and continue to multiply at the same rate as in the preceding experiments.

Table 10.1: Nutrient Requirements of Tetrahymena pyriformis

L-Amino Acids	Vitamins, etc.	Salts
Arginine	Thiamin	$MgSO_4$
Histidine	Riboflavin	$Fe(NH_4)_2(SO_4)_2$
Isoleucine	Nicotinic acid	$MnCl_2$
Leucine	Pyridoxin	$ZnCl_2$
Lysine	Folic acid	$CaCl_2$
Methionine	Pantothenic acid	$CuCl_2$
Phenylalanine	—	$FeCl_3$
Serine	Purine (guanine)	KH_2PO_4
Threonine	Pyrimidine (uracil, cytidine)	K_2HPO_4
Tryptophan		
Valine	α lipoic acid	

Even if food is available, microorganisms may die when other conditions become unfavourable. For example, yeast grown anaerobically produces alcohol and carbon dioxide; alcohol, as it accumulates, eventually becomes toxic to the yeast. Similarly, lactic acid bacteria produce lactic acid in such concentrations that the low pH of the medium kills off all other microorganisms and finally the lactic acid bacteria themselves. Man has made use of some of these special byproducts of microorganisms, e.g., yeast to bake and to make alcoholic beverages

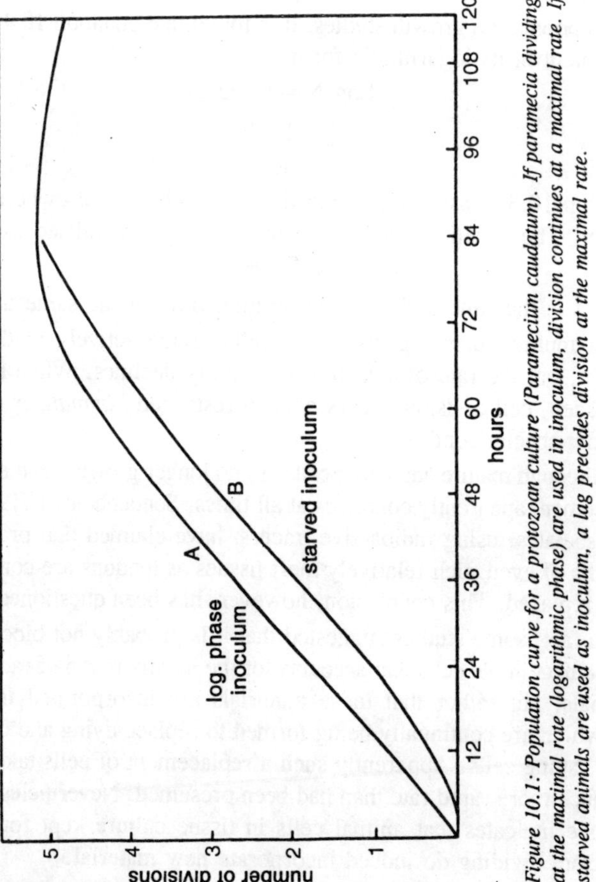

Figure 10.1: Population curve for a protozoan culture (Paramecium caudatum). If paramecia dividing at the maximal rate (logarithmic phase) are used in inoculum, division continues at a maximal rate. If starved animals are used as inoculum, a lag precedes division at the maximal rate.

and lactic acid bacteria to preserve ensilage and cabbage.

If microorganisms did not die for lack of food or because of other unfavourable conditions such as extremes of pH, temperature, radiations, poisons and so forth, or because of the presence of enemies, they would soon exhaust the nutrients of the earth, since their growth is exponental:

$$N = 2^n \quad (10.1)$$

In this equation N is the number of individuals, and n is the number of divisions.

It will be seen that such a growth curve resembles a compound interest curve, because each increment resulting from a cell division doubles the population, which will double again at the next division.

Such an exponential increase in population is often said to be explosive.

In population growth studies, it is found that equation 10.1 is more convenient in its logarithmic form:

$$\text{Log N} = \text{n log 2} \qquad (10.2)$$

or

$$n = \frac{\log N}{\log 2} \qquad (10.3)$$

Equation 10.3 is used in plotting the data in Figure elsewhere in this chapter. The equation may be put into a more generalised form:

$$N = 2^{t/\tau} \qquad (10.4)$$

where t is the time and τ the generation time in the same units.

In multicellular organisms the cells divide actively in the early embryo, but the rate of cell division rapidly declines. When maturity is reached, cell division occurs only in restricted "*immature*" tissues, or under special conditions.

Although mature and aging cells are no longer growing and dividing, replacement apparently continues at all times. Schoenheimer (1942) and his associates, using radioactive tracers, have claimed that proteins in the cells of even such relatively inert tissues as tendons are continually being replaced. This conclusion, however, has been questioned.

In fact, some studies suggested that it is probably not biochemical replacement in old cells that accounts for the incorporation of radioactive materials, but rather that these materials are incorporated into new cells which are continually being formed to replace dying and disintegrating, aging cells. Apparently such a replacement of cells takes place at a much more rapid rate than had been presumed. Nevertheless, some evidence indicates that animal cells in tissue culture kept for a long time from dividing do indeed incorporate new material.

Cells of aging multicellular individuals usually become somewhat dehydrated, respire more slowly, accumulate calcium, various pigments, e.g., pigments in lipid called lipofuscin granules, and other inert matter, and show a decrease in enzymatic activity. The inert matter may be insoluble excretory residues of metabolism which are incapable of passing through a cell membrane. Dehydration of the cytoplasm, resulting perhaps from insolubilisation of proteins and loss of their water-binding capacity, is noted in cells of many but not all species.

It would be of interest in this respect to follow the molecular changes of various cellular constituents with aging of the cell. Such a comparative study has been made with collagen, which forms about 40 per cent of the protein in the human body, being the major constituent in tendon and connective tissue. Experiments indicate that on heating, the tension exerted by collagen fibers from old individuals is greater than that of young individuals and the change is progressive with age. Although the molecular basis of the change in collagen has not been fully established, it appears that crosslinking by ester linkages increases with age, conferring on collagen this particular property.

Why do cells of multicellular organisms age? It could be because the individual cells making up the body have completed their life cycle, or because some unfavourable conditions have arisen. Again, perhaps the body no longer is able to provide adequately for the constituent cells, like an exhausted culture medium for unicellular organisms. Tests of some of these possibilities are considered in the following paragraphs.

MEASURING GROWTH RATES OF CELLS

Growth of individual cells as expressed by increase in mass can be measured using the diver balance, provided the cells are large enough. An *Amoeba* serves such a project well. It is doubtful, however, whether it will ever be possible to study growth in a minute cell such as a bacterium in this manner.

Inasmuch as in a given population cells are generally of the same size, the number of cells at any one time can be determined by direct counts on an aliquot of the culture. Protozoans can be counted directly in aliquots of the medium, even under low magnification.

Bacteria, which are smaller, must be counted under greater magnification than protozoans in a chamber containing a grid and so arranged as to hold a fixed volume of fluid, e.g., the Petroff-Hauser counting chamber. Cells in the blood, eggs of marine animals, protozoans and so forth may also be counted in electronic counters (e.g., the Coulter counter) in which each cell passes through a chamber in which the resistance between two oppositely stationed electrodes changes as the cell passes through and is recorded as a pulse.

A sample can thus be placed in the counter and the total number of cells recorded. It is even possible to count cells of different sizes and to record them separately. A much simpler method which combines the direct count as a calibration with an optical density reading for total numbers has been widely employed for study of growth of populations of small cells such as bacteria.

The optical density of a dense population is first measured and the optical densities of known dilutions of the sample are determined. For a standard curve counts are made of the number of cells in several of the samples for which the optical density is known.

The optical density method is satisfactory, provided growth is normal and cells divide regularly. When abnormal growth occurs, for example, when instead of dividing the cells form filaments, then the optical density may increase linearly even though the genome may not have replicated. The optical density by itself does not, therefore, give sufficient information about growth.

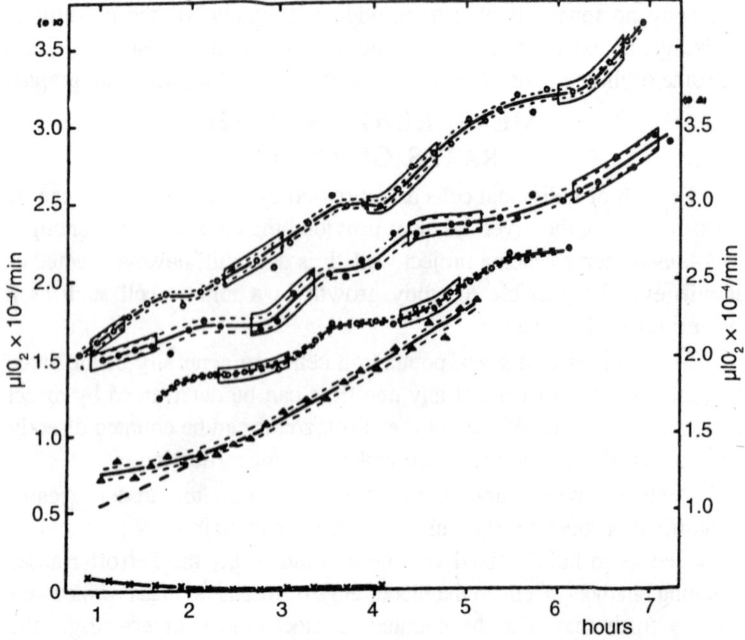

Figure 10.2: Respiration in populations of Tetrahymena cells plotted on arithmetical scales against time in hours. Open and full circles: three experiments with synchronised populations.

Neither counts of cells nor optical density measurements tell how many of the cells recorded are capable of division. The best way to find this is to test a sample of a bacterial culture, for example, on a nutrient agar plate for a colony count. Since a colony is visible only when a relatively large number of cells is present, the presence of a colony indicates an initial viable, healthy cell which has undergone many rounds of replication.

With protozoans it is possible to determine periodically the number of cells in a small test tube (or other container) and so to follow the division rate of a single cell lineage. Because, under strictly controlled environmental conditions, oxygen consumption is proportional to the number of cells, provided an adequate oxygen supply is available, it is possible to determine the growth of cells by the increase in oxygen consumption. The change in total concentration of certain respiratory enzymes in growing cells under an unlimited oxygen supply may also be used as a measure of the rate of growth of cells, but again precautions must be taken to control the environmental factors.

Growth of the amount of certain cell constituents may be followed in living cells by the rate of incorporation of tracers, e.g., ^{35}S incorporated into protein. Since protein increases continuously in a healthy, growing culture, the label concentration in the culture may be a satisfactory measure of the overall rate of growth. Tritiated thymidine and ^{14}C-thymidine have been used in a similar manner to label the DNA.

It must be remembered that replication is semiconservative. Therefore, the ratio of the unlabeled to the labeled chromosomes will become smaller the larger the number of cell generations and replications taking place in labeled phymidine.

GROWTH OF CELLS IN TISSUE CULTURE

Cells in culture, when the cells are transferred periodically to fresh, controlled medium, pose a unique opportunity to study cell growth and aging processes in a variety of cells.

Cells from various plant or animal tissues have been successfully cultured in the absence of other living things. In a typical culture medium for cells of animal tissues, plasma serves as a substrate for attachment of the cells, and an extract from an embryo usually serves as nutrient, although synthetic media are now widely used-see Table elsewhere in this chapter. The cells must be periodically washed free of metabolic wastes.

If these conditions are met, the tissue cells are capable of living, growing and dividing at a constant rate year after year and may be alive and fully vigorous many years after the animal from which the tissue cells were taken would have normally died. Thus Carrel isolated a piece of chick heart and kept it alive in culture from 1913 to 1946, when the experiment was terminated by choice. Plant cells, too, have been grown in tissue culture and in completely synthetic media, and the requirements for different types of plant tissues have been determined.

When cells of human tissues, kept in tissue culture, are dissociated,

Table 10.2. A Serum-Free Medium That Supports Quantitative Growth of Single S3-HeLa Cells

Components	Grams/Liter	Components	Grams/Liter
		1. Amino Acids	
L-Arginine HCl	0.015	L-Valine	0.010
L-Histidine HCl	0.015	L-Glutamic acid	0.030
L-Lysine HCl	0.032	L-Aspartic acid	0.012
L-Tryptophan	0.008	L-Proline	0.010
β-Phenyl-L-alanine	0.010	Glycine	0.040
L-Methionine	0.010	Glutamine	0.080
L-Threonine	0.015	L-Tyrosine	0.016
L-Leucine	0.010	L-Cystine	0.003
L-Isoleucine	0.005		
		2. Vitamins and Growth Factors	
Hypoxanthine	0.010	Choline	0.0012
Thiamine HCl	0.0020	Calcium pantothenate	0.0012
Riboflavin	0.00020	Niacinamide	0.0012
Pyridoxine HCl	0.00020	nositol	0.0004
Folic acid	0.00004	Vitamin B_{12}	0.0010
Biotin	0.00004		
		3. Salts and Other Small Molecules	
NaCl	7.40	$CaCl_2 \cdot 2H_2O$	0.016
KC1	0.285	$NaHCO_3$	1.20
$Na_2HPO_4 \cdot 7H_2O$	0.29	Glucose	1.10
KH_2PO_4	0.083	Phenol red	0.0012
$MgSO_4 \cdot 7H_2O$	0.154		
		4. Proteins	
Normal serum albumin	2.0	Fetuin	2.0

clones can be grown from single cells in colonies much like microorganisms. Such cultures are invaluable for studying many problems. They show that cells of man, e.g., epithelial cells and fibroblasts (cells of connective tissue), like those of the chick, grow rapidly and multiply indefinitely if supplied with the proper nutrients and conditions and washed free of wastes. Protein-free culture media have been developed for this purpose.

GROWTH OF CELLS IN TUMORS

Under certain conditions a cell previously in stasis (not growing or dividing) begins to grow and divide actively, forming *a tumor,* or a cancerous growth. A *cancer* is the result of uncontrolled proliferation of undifferentiated cells when the cells grow at an exponential rate comparable to that of microorganisms, as shown in Figures elsewhere in this chapter. The cancer may remain localised in a given area, or it may *metastasize* (spread by way of the lymph or blood) to all parts of the body.

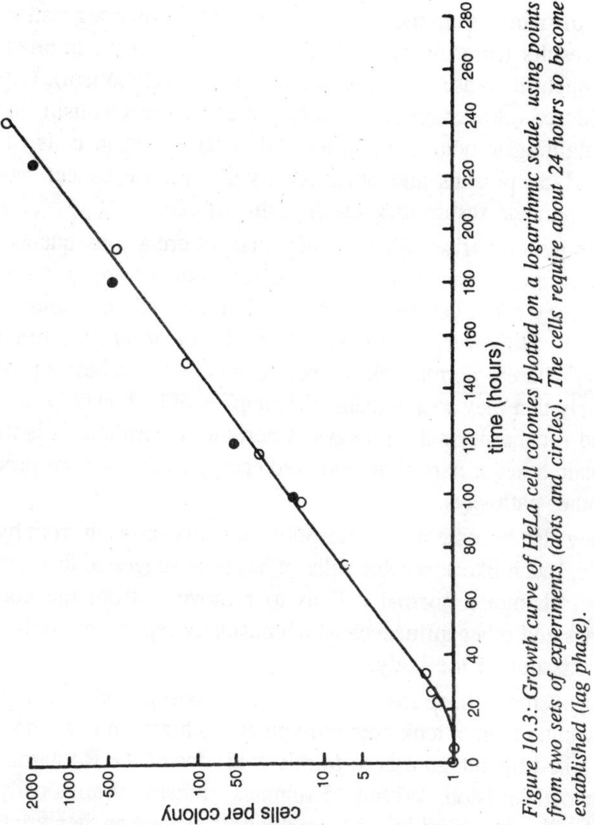

Figure 10.3: Growth curve of HeLa cell colonies plotted on a logarithmic scale, using points from two sets of experiments (dots and circles). The cells require about 24 hours to become established (lag phase).

The undifferentiated cells grow and disorganize the tissues among which they become located. The cancer cells do not appear to be markedly different in nucleolar or nuclear fine structure or in macromolecular syntheses and biochemistry. To prevent or cure cancer it will be necessary to gain a fundamental knowledge of the nature of cell growth and of the mechanism by which the body normally holds the growth and division of cells under control.

At present cancer is treated by various techniques which aim only to suppress its rate of growth once it has started. Most of these techniques originated from study of cell activities. These techniques include ionizing radiations, to suppress division of cells in certain types of cancer; radioactive salts selectively incorporated into a tissue containing cancerous cells, to suppress cell division; and substances which block the synthesis of metabolites necessary for the growth and division of cancerous cells.

On a cellular basis interpretation of cancer induction is still lacking, in spite of many attempts. However, one interesting suggestion is that cancer results from injury of the cellular respiratory equipment, with a consequent increase in glycolysis (anaerobic metabolism). Glycolysis is considered a lower order of energy-liberating mechanism, incapable of maintaining structural integrity and organisation in cells of higher forms. Many poisons and other agents that induce cancer have such effects upon the respiratory mechanisms of cells.

The question arises whether such changes are a consequence, rather than the cause, of cancer. When cells from a tumor are grown in tissue culture, they often behave and respire like normal cells under similar conditions. However, when placed inside an organism, tumor cells multiply, whereas normal cells do not. Tumor cells synthesize protoplasm so actively that they soon exhaust the supplies of high energy phosphates required for the normal pathways of aerobic respiration. It is then that they obtain a major part of the required energy from reactions proceeding along other pathways.

They also become more susceptible to injury or death from hydrogen peroxide, much like anaerobic cells. It has been suggested that a mutation could so change a normal cell as to remove it from the control of hormones and other influences which normally repress multiplication of cells of tissues in the body.

For example, bone marrow cells from a normal individual and bone marrow cells from a leukemic individual, both grown in tissue culture, are inhibited by antagonists to folinic acid, one of the B vitamins required for cell division. Within 15 minutes, division dramatically stops. Normal cells remain inhibited from dividing as long as they are exposed to the antagonist. However, the leukemic cells begin to divide actively after 24 hours, though still exposed to the antagonist. During this time they appear to have inactivated the antagonist because the filtrate from

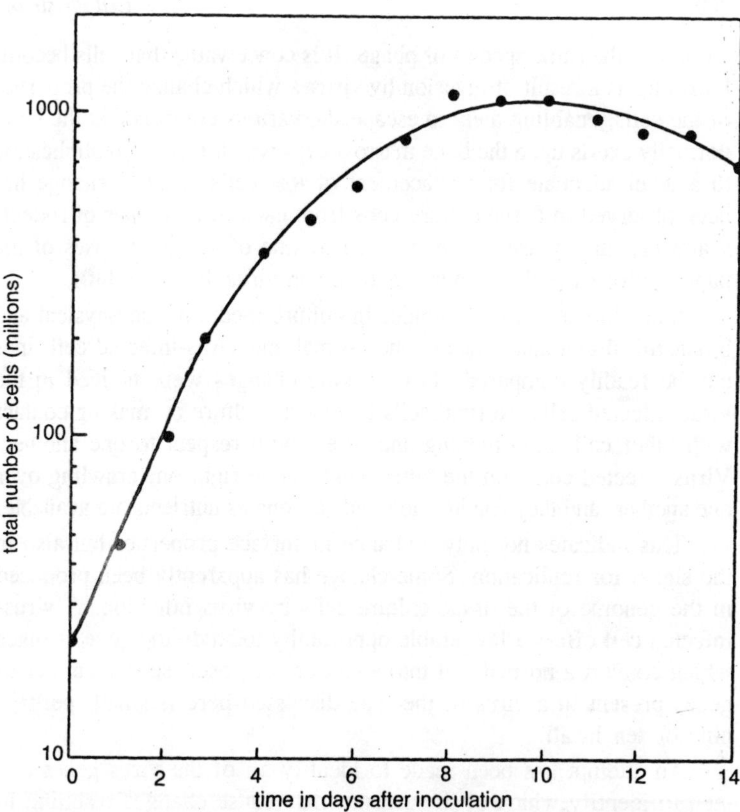

Figure 10.4: Population growth curve for cells of the Ehrlich ascites tumor grown in the mouse. This tumor is made up of individual cells growing in the body (ascitic) fluid.

this culture no longer inhibits either normal cells or leukemic cells, whereas the filtrate from the experiment with the normal cell culture is still inhibitory. (The inactivated material has been separated, crystallised and reactivated.) Apparently, the leukemic cells differ from the normal ones in being able to develop an enzyme system capable of handling the folinic acid antagonist.

Since enzymes are considered to be produced by genes, it would seem likely that the leukemic cells differ genetically from the normal ones.

Another interpretation is also possible. It will be recalled that lysogenic viruses "*hide*" in bacterial cells, remaining inactive except under certain circumstances. But they do produce a change in the host cell since the bacterium is then not infected by the active or lytic variety of the same species of phage. It is conceivable that cells become leukemic as a result of invasion by viruses which change the properties of the cells, enabling them to escape the various controls that the body normally exerts upon the bone marrow cells which restrict multiplication to a level adequate for replacement of lost cells. Such a change has been observed in tissue culture cells from each of a number of rodents (hamster, rat, mouse) when infected by one of several viruses of the papova viruses (polyoma virus and simian virus 40, or SM40).

Since the cells form colonies in culture medium, the physical and biochemical characteristics of the normal and virus-infected cell lines can be readily compared. Two striking changes were noticed in the virus-infected cells. Normal cells in a tissue culture on making contact with other cells stop moving and orient with respect to one another. Virus-infected cells, on the other hand, move right on, crawling over one another, and they continue to divide as long as nutrients are available.

This indicates not only a change in surface properties but also in the signal for replication. Some change has apparently been produced in the genome of the tissue culture cells by virus infection. A virus-infected cell offers a favourable opportunity to study the gene changes which convert a normal cell into a cancer cell, because the number of genes present in a virus of the type discussed here is small, perhaps nine or ten in all.

An attempt has been made to identify all of the virus genes and then to identify which of these genes may cause changes resulting in cancer induction. In the polyoma virus there is enough DNA (about 5000 base pairs) to code for about eight or nine proteins. Probably one or two code for the protein in the viral coat. Six functions, each arising in the host cell after infection and each coded by one gene in the virus, have now been studied in the infected cells.

These functions involve changes in biochemical or immunological properties and probably depend upon proteins produced by these genes. None of these functions have been specifically related to the two cancerous properties which were observed in virus-infected tissue culture cells-contact inhibition and uncontrolled multiplication. It is recognised that mutation, without virus infection, may occur in a cell of an animal (or plant). Such a mutation may result in a change to a cancerous condition, releasing the cell from regulation which normally controls cell division. However, it is difficult to identify such a genetic change because there are so many genes in the genome of the animal cell that analysis is made immensely more complicated.

Reversal of the tumorous condition of cells in plants and animals has raised the question whether the change in some cases might not be an altered expression of a gene rather than a change in genome. Thus the crown gall tumor can be reversed by a series of transfers to healthy plants.

Reversion of tumors has also been observed in both newt and mouse. A variety of evidence for viral and for genetic origins of cancer has accumulated, none yet permitting a successful analysis of the basic change involved. Cancer remains an important problem to the cell physiologist.

REGULATION OF FORM AND SIZE IN CELLS

The ability of a cell to maintain its organisation and characteristic size at a given temperature and in the presence of adequate nutrients is perhaps best shown by the regeneration of protozoans after removal of parts of a cell.

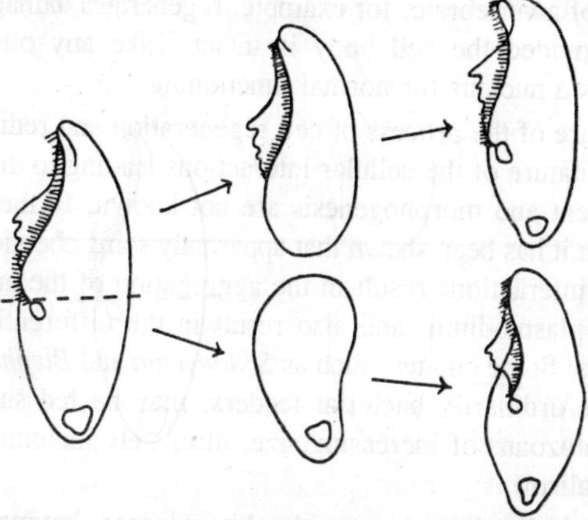

Figure 10.5: Regeneration of missing parts in two pieces of a bisected Blepharisma. *Note especially the regeneration of the hypostome (mouth region) in the posterior segment. Regeneration occurs within about 6 hours at room temperature.*

In fact, *a Blepharisma* from which the entire hypostome (mouth area) has been removed regenerates it sufficiently well to feed within 6 hours. If large portions of the cell are removed, but macronuclear fragments and micronuclei remain, regeneration will result in a miniature edition of the species. After feeding, however, the regenerated organism rapidly regains its original size. Redifferentiation is usually restricted to portions of the cell adjoining those removed by microsurgery; however, even then, materials required for regeneration must be obtained from existent structures. In some species of ciliates, amazingly enough, even a small injury results in dedifferentiation of the entire animal and redifferentiation of all the organelles (such as membranelles, motile organs and mouth parts). *Stentor* mangled into about 40 adherent pieces survives and regenerates its normal form .

As a result of this morphog-enetic activity, organelles which have been removed or damaged are regenerated, and the characteristic form and size of the species are reconstituted. When pieces are removed from eggs of some chordates and echinoderms, the eggs produce complete embryos of reduced size. This shows that the egg maintains a certain organisation, just as does the unicellular organism. When pieces are removed from the eggs of annelids or mollusks, the eggs develop into partial embryos only.

In such cases the pattern is supposedly already laid down earlier in these eggs and each part is already destined to form a particular organ. It is probable that most vertebrate cells can recover from minor injuries, although experimental studies on this subject have been limited. The neuron of a vertebrate, for example, regenerates damaged axons or dendrites provided the cell body is intact. Like any other cell, the neuron needs a nucleus for normal functioning.

The nature of the process of cell regeneration and redifferentiation and also the nature of the cellular interactions leading to differentiation in development and morphogenesis are not known. In the slime mold *Dictyostelium* it has been shown that apparently some chemical attraction and cellular interactions result in the aggregation of the ameboid cells to form the plasmodium, and also result in the differentiation of the fruiting stalks. Some ciliates, such as *Stylonychia* and *Blepharisma,* both of which are ordinarily bacterial feeders, may be fed successively a series of protozoans of increasing size, ultimately including their own kind (cannibalism).

As a result they become considerably enlarged, having many times the bulk of the normal bacteria-fed individuals. The mouth parts, membranes, motile organs, macronuclei and other structures become proportionally enlarged, and the large size is maintained so long as the abnormal diet is continued. In all cases, however, it has been noticed that when the giants are again put on a bacterial diet, division is more rapid than it is in controls fed continuously on bacteria, so that the giants soon return to the characteristic species size.

Conversely, ciliates may be starved for long periods, during which they undergo reduction in size. *Tetrahymena,* for example, will become a minute particle with little resemblance to its former self. Yet on addition of nutrients it quickly enlarges to the normal size, and when it has grown double this size it divides, thus maintaining the normal size of the species. That size of cells is regulated by division is most clearly demonstrated with *Amoeba proteus* . During growth of an ameba the mass increases until it just doubles.

Then no further increase occurs, and after a lapse of time it divides, each daughter cell receiving half the mass of the mother cell. If a part of an ameba is cut off during the period of growth, division will not occur at the time it does in control amebas, but only after the removed mass of protoplasm has been replaced by growth.

An ameba may be prevented from dividing, presumably indefinitely, by the periodic removal of a piece of protoplasm, although there is some evidence to the contrary. Heredity plays an important role in the maintenance of size and shape of cells. This is evident since clones of a given species of microorganisms under a given set of environmental conditions tend to be quite constant in size and form for a long series of generations-a constant nuclear to cytoplasmic ratio is maintained. However, overall size was found to vary with a change in each environmental factor studied. Experiments have shown that increasing the number of chromosomes by polyploidy is apparently often attended with an increase in cell size.

If the number of chromosomes is increased, each chromosome maintains its characteristic size; therefore the amount of chromatin present in the polyploid nucleus is correspondingly increased. The nucleus increases proportionally in size, and the cytoplasm compensates. A polyploid salamander, for example, has giant cells. The same is true of the cells in certain mosses which are usually haploid but may be experimentally produced in diploid, triploid or tetraploid form. Polyploidy wherever found is always correlated with increased size of the polyploid cell.

In plant cells, growth might appear to be limited by the rigidity of the cell walls. However, the cell wall is known to be rendered softer and plastic by the plant growth hormone, auxin (indole acetic acid), this being, presumably, what occurs to cells in the region of growth. For example, when plants illuminated on one side only turn toward the light, auxin was shown to accumulate on the side away from the light, inducing extra elongation of young cells in the region of growth on that side.

The curvature is thus a result of increase in cell size and not from cell multiplication. Even in the dark application of auxin to one side of a plant results in curvature from increase in cell size on that side. Similar softening of cell walls occurs whenever plant cells increase in volume after cell division, but here the action of auxin is less evident than in tropistic responses. Again, during embryogeny plant cells lack a central sap vacuole. Differentiation usually involves the development of a large central sap vacuole, along with the resultant enlargement of the cell. During such increases in cell size, auxins render the cell walls plastic. It is conceivable that regulation of growth in plant cells may be controlled in part by the auxin supply.

UNBALANCED GROWTH

When a mutant form of *E. coil*, which requires for its growth the pyrimidine base thymine, is grown in a medium deficient in thymine but adequate in all the other nutrients, growth (as measured by turbidity) apparently occurs at essentially the normal rate. However, when this culture has been grown in the deficient medium for a period equivalent to that required for one generation and then is plated out, even on a balanced medium containing all necessary nutrients, including thymine, it is observed that about 90 per cent of the individuals fail to produce colonies.

Injury therefore appears to occur when replication is attempted. If so grown for two generations, about 99 per cent fail to form colonies. Tests show that protein and RNA increase in the cells, but there is only a slight increase in DNA. Such disproportionate growth of different cell constituents is spoken of as *unbalanced growth*. A similar phenomenon has been observed in yeast and *Neurospora*.

However, it does not appear to be general, since some cells recover even after becoming much enlarged. Recent studies on repair mechanisms in cells have indicated that during the course of replication some breaks occur in the double-stranded DNA as it opens before the growing point. Such breaks occur, perhaps at a relatively high frequency, and are apparently readily repaired provided the nucleoside triphosphates and the necessary enzymes are present. For example, if thymine is withdrawn from the medium of a thymineless mutant of *E. coli* and DNA replication is attempted, the DNA molecule breaks and cannot be repaired because of the lack of the nucleoside containing thymine. Consequently the cell cannot replicate, the chromosome fragments, and the cell dies. A similar course of events may occur in other cases as well.

An interesting experiment is one in which a mutant requiring some nutrient common to synthesis of both RNA and DNA is maintained alive on a medium deficient in the nutrient. In this case, no growth, or very little, can occur. Presumably it is only when growth is unbalanced, that is, when most constituents can increase in quantity but one does not, that injury occurs.

It might be expected that any factor (other than limitation of nutrients) which leads to unbalanced growth might injure the cell. Experiments with ultraviolet radiations (to which synthesis of DNA is one of the most susceptible processes) show that the irradiated cells in which DNA synthesis is just inhibited may continue to grow (as measured by optical density) and individual cells become, giants accumulating RNA and proteins.

Plated out, however, they also fail to produce colonies-i.e., they are unable to divide. It is argued that the unbalanced growth of various structures in the absence of DNA synthesis has deranged the cell so that it is no longer able to divide and reproduce, even when later the DNA synthesis has apparently recovered from the radiation injury. If, however, the cells are irradiated sufficiently to stop DNA synthesis and then kept in non-nutrient solution so that they cannot grow, DNA synthesis recovers slowly, and addition of required nutrients will restore balanced growth. These are the very conditions under which dark repair can best occur.

REGULATION OF GROWTH

Balanced growth is a steady state condition in which every component of the cell doubles in a cell generation (time between divisions). The generation time depends upon the species and in part upon the conditions of the environment. Thus, *E. coli,* on a medium containing water, glucose, ammonium chloride, potassium basic phosphate, sodium sulfate and trace metals, kept at 37° C., divides every 45 minutes. If supplied with amino acids and some other organic materials at 37° C., it divides every 20 minutes. The cells have thus reached a new steady state in which duplication of cell constituents occurs much quicker. Division is considerably retarded at 15° C., at which temperature a cell gives rise to only about 128 cells after 24 hours, whereas at 37° C. the number would be about 2×10^9.

Cell growth remains a challenging problem for experimental study. The question at hand is to determine how a cell achieves balanced growth regulation. Since growth is a result of metabolic activity along various fronts, the questions really asked are: how is metabolic activity regulated? and how are the various types of metabolic activity coordinated?

The induction of enzyme synthesis and the repression of enzyme synthesis have already been discussed, and it will be recalled that the end product inhibition is a means of regulating many of the steps in synthesis, producing immediate changes in rate of reactions and thus preventing accumulation of cell products. In Figure elsewher in this chapter the interplay between the synthesis and end product inhibition of various substances needed for the synthesis of DNA and RNA is considered.

The net effect is a proportionality between synthesis of the two important nucleic acids. A similar interplay in other synthetic pathways perhaps leads to production of useful amounts of proteins and other substances needed by the cell. Thus a balance may be achieved in the amounts synthesized of the various requirements for cell growth. It is likely that steady state necessary for balanced growth is thus achieved.

11

Transport Through Cell Membrane

Many substances other than water move in and out of cells in response to a concentration gradient, although only a few of them (carbon dioxide and oxygen) are able to move as freely as water. If a cell membrane allows a substance to move in or out through it, it is said to be permeable to that substance.

METHODS FOR STUDYING PERMEABILITY OF CELL MEMBRANES TO SOLUTES

Only a few of the methods employed in permeability studies will be discussed here. Some methods are only qualitative. For example, entry of dyes, such as neutral red, into cells like paramecia can be observed by the appearance of a pink colour inside the cells. Sometimes use is made of such dye penetration as an indicator. Thus the entry of a base like ammonia into paramecia stained with neutral red may be observed because the pink paramecia become yellowish as soon as the base has entered. However, it is virtually impossible to make quantitative measurements because if the penetration is slow, or if only a minute quantity enters, no colour change is noticed. Furthermore, the method depends upon a personal element judgment of colour—which is likely to vary from one person to another. The same criticism applies also to other visual methods, such as the formation of an insoluble precipitate inside cells (e.g., the formation of calcium oxalate crystals in cells of *Elodea* as an indication of the penetration of oxalic acid).

Volume Change

A quantitative method much more satisfactory for studying the permeability of cell membranes is the observation of the rate of deplasmolysis of plasmolysed cells. The relative rates of entry into cells of different substances used in hypertonic solutions may be tested by a study of rates of deplasmolysis as the cells reabsorb water. For example, when equimolal hypertonic solutions of several substances like ethylene glycol, glycerol or glucose are applied to cells, they will plasmolyze the cells. Deplasmolysis will occur first among the cells in ethylene glycol and last among those in glucose. This means that ethylene glycol penetrates the fastest of the three and glucose the slowest. If the cell remains shrunken in a solution, the solute has failed to enter the cell or changes in the cell structure have occurred.

A similar method, particularly useful for the study of the permeability of the plasma membranes of minute cells, is the determination of the volume of a mass of cells before and after immersion in hypertonic solutions. The volume of the cell mass is determined after the cells have been packed together by centrifugation. This is a favourite method for the study of permeability of red blood corpuscles.

A special centrifuge tube called the *hematocrit*, with a calibrated fine cylindrical portion 1 or 2 mm. in diameter, is used for this purpose. If the series of experiments with different substances is carried out under similar conditions and with the same centrifugal force applied for the same time, quantitative data on the rate of entry of the various substances can be readily obtained by the volume changes of the packed cells. However, it is virtually impossible to pack the cells without any intercellular fluid; therefore such experiments should be designed to compare relative packing of a control and experimental sample.

Hemolysis

Hemolysis, indicated by the sudden appearance of a clear red solution instead of the previously murky suspension of red blood cells, has also been used as a measure of the rate of penetration of substances into erythrocytes. For this purpose red blood cells are placed in hypertonic solutions such as ethylene glycol. At first water leaves the cells, causing crenation, but in time ethylene glycol molecules enter. This entry increases the store of osmotically active molecules inside the cells, which results in a corresponding uptake of water.

The cell membrane can stretch only to a limited extent without altering its properties, however, and the cells undergo *hemolysis* when this limit has been passed; that is, the membrane has become porous enough to permit the hemoglobin molecules to leave the cell. In solutions of glucose, erythrocytes are hemolysed only after a much longer time than when they are in ethylene glycol.

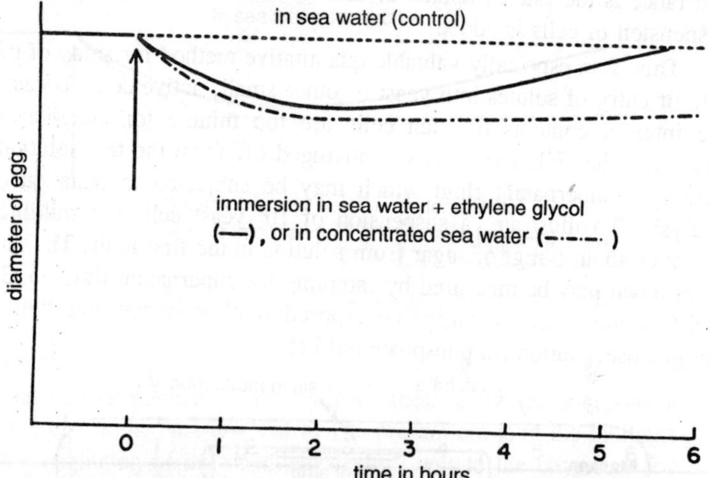

Figure 11.1: Return to normal volume as a measure of permeability.

The relative rates of entry of various kinds of molecules may therefore be measured by the rates of hemolysis. However, some question exists whether this method can be used for comparisons of hemolysis of red blood cells in different species, because the strength of their membranes may vary and thus hemolysis rates may not have the same meaning.

Quantitative Analyses

Perhaps the most direct method for measuring permeability of cells is the quantitative analysis of the internal contents of a cell exposed to a solute for a known period. For this purpose large coenocytic (multinucleate) plant cells, such as *Valonia, Chara* and *Halicystis,* are ideal, because the uptake of the materials can be measured by analysis of the contents of the large central sap vacuoles.

The vacuole of *Valonia* or *Halicystis* often contains more than 1 ml. of sap, making possible microchemical tests on a single cell. Few animal cells are large enough to permit chemical analyses of a single cell, but the squid giant axon has been so employed. The chemical method is especially valuable because the data lend themselves to calculation of permeability constants in terms of the quantity of material transported per unit time across a unit area of cell membrane per unit osmotic gradient. Data obtained on one cell for a variety of substances may therefore be quantitatively compared with data on another cell. Still another method used to measure permeability of a cell to a given substance is the rate of uptake of that substance from solution after a suspension of cells is added.

This is an especially valuable quantitative method for study of the rate of entry of solutes into yeast or other small, active cells, because the internal contents of such cells are too minute for quantitative analysis. The cells are readily centrifuged off from the test solution, leaving a supèrnatant fluid which may be subjected to quantitative analysis. To illustrate, a suspension of 10^8 yeast cells per milliliter removes about 6 mg. of sugar from solution in the first hour. The rate of removal may be measured by sampling the supernatant fluid 15, 30 and 60 minutes after yeast is first exposed to a known concentration of the glucose solution (in phosphate buffer).

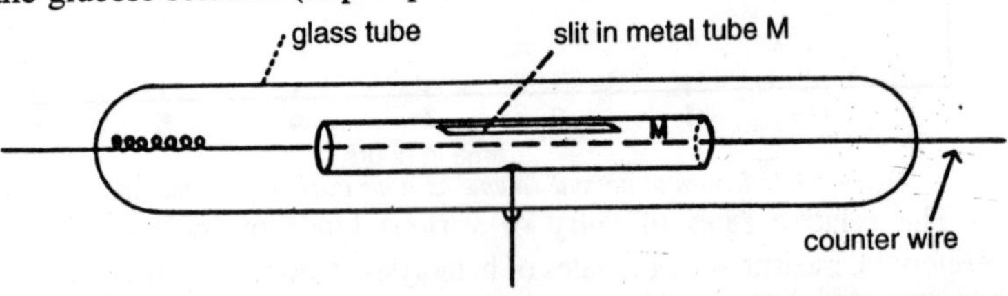

Figure 11.2: A Geiger-Muller counter (diagrammatic). This is registered as a "plunk" by the relay-driven counter, which in simplest form resembles a cyclometer or tally counter.

A small correction has to be made for the glucose respired during the course of the experiment. Another possible source of error is adsorption of some solutes to the surfaces of the cells (classroom experiment). Another important quantitative method is that employing radioactive tracers. Cells may be exposed for a given period to solutions containing radioactive isotopes, e.g., ^{24}Na.

After the test suspension is centrifuged, the supernatant is removed and the cells are rinsed off. The amount of radioactive ion which has been taken up may be measured with a Geiger counter. By making readings after varying the exposure time to radioactive sodium, one may determine the rate of ion uptake. Moreover, periodic readings taken after the cells have been removed from the radioactive solutions and suspended in nonradioactive ("cold") medium, indicate the rate of loss of ions.

Valuable quantitative data have been obtained by this method. For instance, unique data not obtainable by other methods have thus been gathered for sodium. Previous experiments suggested that sodium does not enter cells, or enters so slowly that it cannot be measured. Exposure of cells to radioactive sodium, however, demonstrated that sodium ions enter and leave cells at the same readily measurable rate; therefore, no net change occurs in the cell content of sodium ions. Consequently, movement of sodium ions could not have been demonstrated by even the most precise quantitative chemical determinations.

The radioactive isotope method has been confined largely to study of the entry of a number of electrolytes which have radioactive isotopes with sufficiently long half-life periods to permit measurement. ^{3}H, ^{14}C, ^{24}Na, ^{32}P, ^{35}S, ^{36}Cl, ^{42}K, ^{45}Ca, ^{59}Fe, ^{64}Cu, ^{65}Zn, ^{82}Br, ^{89}Sr and ^{131}I are some of the isotopes available for tracer studies. However, the tracer method need not be confined to electrolytes, since organic molecules may be labeled with radioactive carbon, etc..

THE CORRELATION OF PERMEABILITY WITH PARTITION COEFFICIENTS

A definite rationale lies behind the groups of compounds tested in permeability studies. By testing various groups of compounds, each with certain properties in common, it has been possible to find some correlation between these properties and movements of each solute through the cell membrane.

Nonpolar compounds are compounds in which electrons are equally shared by the two atoms forming a bond, as in the paraffins, olefins and cyclic compounds. They have a high solubility in fats or fat solvents, but a low solubility in water.

In *nonionic polar compounds* the electrons of the bond between two atoms are attracted more to one of the atoms (e.g., to the strongly electronegative atoms, such as oxygen, nitrogen and chlorine) than to the other. As a consequence, the molecule acts as a dipole and orients in an electric field. Nonionic polar compounds are derivatives of nonpolar compounds into which electronegative atoms or radicals have been introduced. Thus, methyl alcohol is a derivative of methane, ethyl alcohol a derivative of ethane, and so on.

In *ionic compounds* the electrons of a valency bond are held by only one of the atoms or radicals, so that ions are present even in the crystalline state of such compounds. In the crystal the ions are bound together by electrostatic forces. A liquid of high dielectric constant, like water, stabilizes the attractive forces between pairs of ions, allowing them to separate from one another and act as individual particles. Ionic compounds are of two *kinds-weak electrolytes,* such as weak acids and bases, and *strong electrolytes,* such as salts and strong acids and bases.

```
    H              H H              H H H
  H:C:H          H:C:C:H          H:C:C:C:H
    H              H H              H H H
  methane          ethane           propane

  H H                Cl             H :O:
H:C:C :O: H      Cl: C :Cl        H:C:C:H
  H H                Cl             H
ethyl alcohol  carbon tetrachloride  acetaldehyde

    :O:               H
 :O:S:O:          H : N: H          H :O: H
    :O:               H                H
   SO4-             NH4+              H3O+
 sulfate ion    ammonium ion     hydronium ion
```

Figure 11.3: Electron configuration in nonpolar and polar compounds. Note that in polar compounds, both nonionic and ionic, the electrons are closer to the electronegative atoms.

The stronger the electrolyte, the greater will be its solubility in water and the less its solubility in oil and oil solvents. In a homologous series such as the organic acids, the strength (ionisation) of the acids decreases as the chain length increases; for instance, formic acid is stronger than acetic, the latter is stronger than propionic, and so on. As the degree of ionisation declines, the oil solubility increases, and the higher fatty acids are freely soluble in oils or oil solvents but only slightly soluble in water.

The ratio of solubility of a compound in oil or fat solvent to its solubility in water is called the *partition coefficient*. The more nonpolar a compound is, the more soluble it is in fats or fat solvents, i.e., the more *lipophilic* it is. Such compounds usually have slight, if any, solubility in water.

Their partition coefficients are therefore high. On the other hand, polar derivatives of nonpolar compounds are usually less soluble in fats or fat solvents, in correspondence to their degree of polarity. An increase in polarity results in greater solubility in water. To illustrate, lower alcohols like methyl alcohol and ethyl alcohol are highly polar, and they are only slightly soluble in fats or fat solvents but completely miscible with water. Conversely, higher alcohols like butyl alcohol and amyl alcohol are sparingly soluble in water but miscible with fat solvents in all proportions. Lower alcohols have low partition coefficients; higher alcohols have high ones.

Ionised compounds are polar and correspondingly less soluble in fats but usually highly soluble in water. Weak electrolytes with lipophilic radicals, like the long chains of the higher fatty acids mentioned before, are soluble in fats and have high partition coefficients, whereas the lower fatty acids (e.g., formic acid, acetic acid) are highly soluble in water and have low partition coefficients. Strong electrolytes are sparingly soluble in fats but highly soluble in water. Their partition coefficients are therefore low.

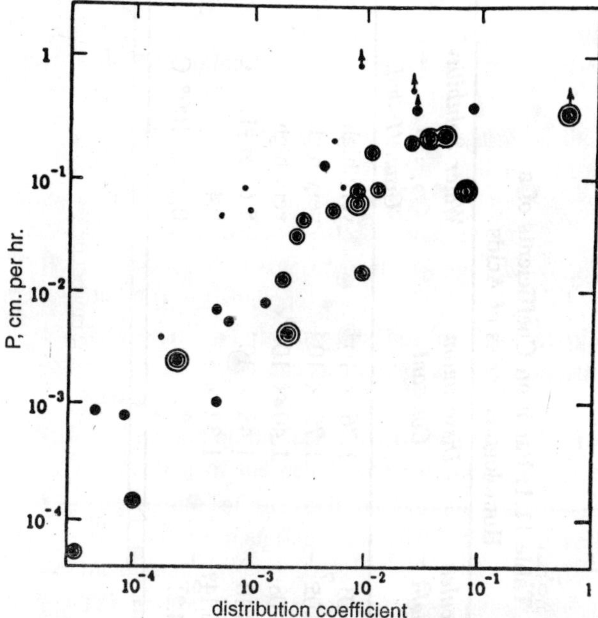

Figure 11.4: Relation between partition coefficient (distribution coefficient) of a substance between oil and water and its penetration (P) into the Chara *cell.*

The entry of many compounds into the large cells of *Chara* was studied by Collander and Barlund, and some of their results are cited in detail here because they constitute the most extensive study of its kind. The amount of a substance entering *a Chara* cell per unit time was determined by chemical analysis, and the permeability constant in terms of the quantity of compound entering per unit time per unit area of cell surface per unit osmotic gradient was then calculated. These data are given graphically in Figure elsewhere in this chapter. The approximate molecular diameters of the compounds are indicated by the size of the circles in the figure elsewhere in this chapter, the largest circles designating the largest molecules.

Scrutiny of Figure elsewhere in this chapter discloses that usually

Table 11.1: Partition Coefficients of a Homologous Series of Acids

Acid	*Number of Carbons*	*Moleculea Weight*	*Dissociation Constant*	*Water Solubility in Grams/100ml.*	*Partitition Coefficient, Benzene : Water*
Formic	1	46.03	1.76×10^{-4}	very high	—
Acetic	2	60.05	1.75×10^{-5}	very high	0.055
Propionic	3	74.08	1.40×10^{-5}	very high	0.28
Butyric	4	88.10	1.48×10^{-5}	5.62 at -1.1 °C	1.08
Caproic	6	116.14	1.32×10^{-5}	0.4	3.73
Caprylic	8	144.21	—	0.25 at 100° C	16.10

molecules with the largest partition coefficients enter the cell most readily, regardless of molecular size. Thus if in a homologous series such as the alcohols the narcotic action of the alcohols on respiration is used to measure their penetration, higher alcohols with greater oil solubility are observed to enter cells much more readily than lower alcohols of less oil solubility, even though the molecular sizes of the higher alcohols are considerably larger. Another example showing that entry of molecules into cells increases with increase in partition coefficient, even though there is an increase in molecular weight, is that of the urethans.

Comparing the effects of urethans (alkyl carbamates) on the resting potential of frog nerves as a measure of their penetration into cells, one can demonstrate that urethan (ethyl carbamate) penetrates less readily than the substituted urethans in which methyl, ethyl or other radicals are substituted for some of the hydrogen atoms. The substituted urethans have higher oil to water partition coefficients than ethyl carbamate.

Table 11.2: Partition Coefficients (Ratio of Solubility in Oil to Solubility in Water) of a Series of Homologous Alcohols

Alcohol	*Molecular Weight*	*Solubility in gm. Per 100 ml. Water*	*Partition Coefficient, Oil : Water*	*"Narcotic Concentration"*
Methyl	31.06	00	0.0097	5.0
Ethyl	46.07	00	0.0357	1.6
Propyl	60.09	00	0.156	0.8
n-Butyl	74.12	7.9	0.588	0.15
i-Amyl	88.15	slight	2.13	0.045

Since the lipoid solubility of a molecule is a factor of greater moment than its size in determining rate of entry into a cell, the question arises whether partition *coefficient* alone determines if a molecule shall pass through the plasma membrane. However, further experiments demonstrate that molecular size does influence the passage of substances through the plasma membrane and that many other factors, such as hydrogen ion concentration, salts, temperature and physiological state, also influence the permeability of the plasma membrane.

THE EFFECTS OF MOLECULAR SIZE UPON PERMEABILITY

In studying the effects of molecular size on permeability, one should note that in general the greater the molecular weight, the larger will be the molecule. Molecular diameters also depend upon steric configuration, especially in organic molecules. Symmetry, or its lack, also influences the effective diameter of a molecule, an asymmetrical molecule having a larger effective diameter than a symmetrical one.

(It is understood that very large molecules probably have only a slight chance of entering a cell through the plasma membrane. Sucrose, for example, penetrates cells in very small amounts, if at all. Larger carbohydrate molecules, such as starch, glycogen or inulin, have not been shown to diffuse through plasma membranes; for this reason they serve effectively as food stores. The probability that a protein molecule will pass through a cell membrane is also very slight, although the absorption of antigens indicates that penetration occurs in some instances, perhaps by pinocytosis.

On the other hand, it has been found that, provided the partition coefficients of the compounds are the same, the molecular size has some effect upon permeability. For example, cyanamide, propionamide, succinamide and diethylmalonamide have about the same partition coefficients, but their molecular diameters increase in the order given. When these compounds were tested on *Chara* it was found that their ability to penetrate the cell membrane decreased with increase in molecular diameter. Size, then, is also a factor, perhaps secondary, in determining whether or not a molecule will pass into a cell and in determining the rate of penetration.

However, there are cases in which molecular size of small molecules may be of even greater importance than the partition coefficient in determining the penetration of molecules into some cells. This is illustrated by the data in Figure elsewhere in this chapter for *Beggiatoa,* a large sulfur bacterium, tested with compounds similar to those used with *Chara.* Here the correlation between permeation and molecular size is much more decisive than the correlation between permeation and partition coefficient.

Another example of the influence of molecular size on permeability is shown in the penetration of aldehydes into a dark mutant strain of luminous bacteria, which lacks the ability to form its own aldehydes needed for luminescence, but will luminesce if aldehyde is supplied. By using the first phase of luminescence (when luminous intensity increases linearly with the concentration of aldehyde) as a measure of entry, the data for rate of penetration of a homologous series of aldehydes was obtained, as shown in Table elsewhere in this chapter. The permeability

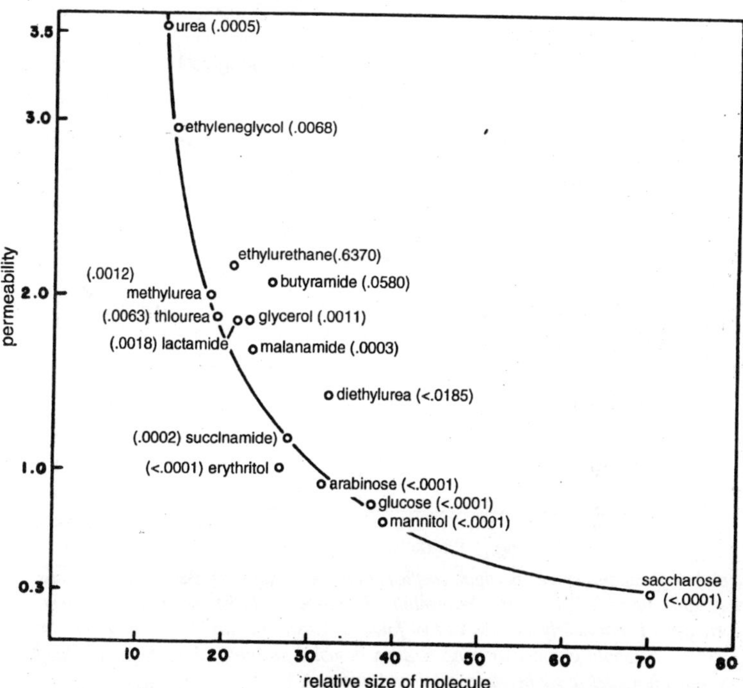

Figure 11.5: Relation between molecular size and permeability in Beggiatoa. Molecular size is measured by molecular refraction in yellow light.

constants calculated for the rate of entry of aldehydes into bacteria are of the same order of magnitude as for entry of other nonelectrolytes into plant cells.

It is apparent that it is impossible for macromolecules such as polypeptides, proteins and polysaccharides to pass through the pores in (or dissolve into) a plasma membrane through which smaller molecules enter and leave the cell ecause of their large size. The cell has developed another method or uptake of such macromolecules, i.e., *pinocytosis,* which is discussed in other chapter of this book.

THE EFFECT OF IONISATION UPON PERMEABILITY

Electrolytes enter cells more slowly, in general, than nonelectrolytes of similar molecular dimensions, and weaker electrolytes enter more rapidly than strong ones. The relation between charge and penetration is best seen by studying the rate of entry of weak electrolytes as affected by their degree of ionisation. The degree of ionisation of a weak electrolyte is determined by the pH. By varying the pH, within

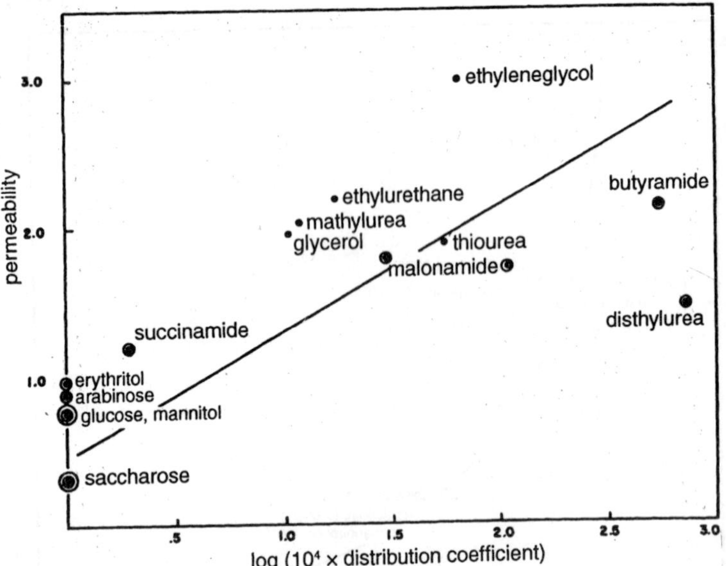

Figure 11.6: Relation between partition coefficient and permeability of Beggiatoa *to various organic substances. Although the permeability increases with the partition coefficient, the correlation is not nearly so good as that found in* Chara. *Permeability is measured by* log_{10} *(10^4 × threshold plasmolytic concentration in moles per liter). The abscissa is* log_{10} *(10^4 × partition coefficient between ether and water).*

limits safe for the living cell, one may determine what effect the charge on an electrolyte has upon its ability to penetrate a plasma membrane.

In Figure elswhere in this chapter is shown the effect of pH on the penetration of carbon dioxide or carbonic acid. At pH 6.34 one-half is dissociated. As the pH rises, the dissociation increases but the entry of carbon dioxide decreases.

Conversely, with fall in pH the amount of dissociation decreases and the entry of carbon dioxide into the cell increases. Similar relations hold for hydrogen sulfide, hydrocyanic acid, auxins (plant growth hormones), dinitrophenol (metabolic stimulant) and other biologically active weak acids.

The entry of weak bases such as the amines and alkaloids is governed by the same principles, except that in their case the rise in pH suppresses dissociation and enhances penetration. The effect of degree of ionisation on the entry of electrolytes into cells is further illustrated by comparing the rates of entry of members of a homologous series such as the weak organic acids illustrated in Table elsewhere in this chpater. As judged by criteria such as colour changes in cells with natural pigments, penetration of the anions was shown to increase in the order given in the table. It appears that in spite of the increase in molecular size, the entry of the acids is more rapid, the smaller their degree of dissociation.

However, lipoid solubility plays a role here also, since the rate of entry increases with a rise in the partition coefficient, a factor previously emphasized as of paramount importance. Presumably, permeation is proportional to the number of undissociated molecules, and a change in pH in the direction which increases the proportion of undissociated molecules will enhance penetration. The data indicate that the presence of a charge on an ion decreases its chance of entry.

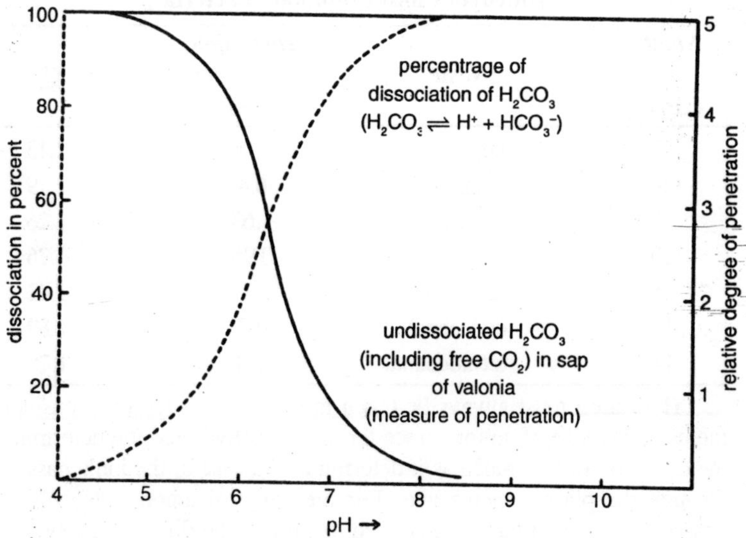

Figure 11.7: Relation between pH and penetration of a weak electrolyte in Valonia.

If the presence of an electric charge is indeed a disadvantage for entry of ions into the cell, [hen the stronger the charge on an ion the less probable should be the ion's entry into a cell. This is found to be true. In studies on a series of strong electrolytes (completely dissociated) some ions are found to enter more readily than others. Monovalent cdtions such as Na^+ and K^+ enter more readily than divalent cations such as Ca^{++} or Mg^{++}, and divalent cations enter more readily than trivalent cations such as Fe^{+++})Similarly, monovalent anions such as I^- or Cl^- enter more readily than divalent anions such as SO_4^- or tartrate.

Not all ions of the same valency enter cells at the same rate. For example, the ammonium ion enters more rapidly than potassium, potassium more rapidly than sodium, and sodium more rapidly than lithium. This might be explained on the basis of the relative size of the hydrated ions. As was explained in other chapter of the book, ions bear charges which attract and orient water molecules about themselves in hydration spheres. The resultant particle is therefore much larger than the ion itself. Hydration is not equal in all ions. For instance, the nucleus of potassium is surrounded by several shells of electrons; therefore its net charge density is lower than that of lithium, which has one shell of electrons. Consequently, less water is attracted to potassium than to lithium.

Table 11.3: Rate of Penetration of Aldehydes into Luminous Bacteria

Number of Carbons	*Name*	*Penetration in cm. Per Sec. × 10^{-6}*	Q_{10}
7	heptanal	10.6	1.13
8	octanal	6.4	1.19
9	nonanal	1.68	1.26
10	decanal	0.95	1.26
11	undecanal	0.85	1.42
12	dodecanal	0.7	1.89
14	tetradecanal	0.31	1.78

Hydration must always be taken into account when one considers the effective size of an ion, since the size of a hydrated ion determines the diffusion rate as much as it determines the ease of the ion's passage through the plasma membrane. For the entry of anions into cells, a similar explanation has been applied to explain the differential entry of ions of the same valency, e.g., SCN I > NO_3 > Cl. Thus permeability to ions is dependent to some extent upon a "*sieve*" action of the plasma membrane. However, it can be demonstrated that the hydration sphere does not electrically neutralize the ion. The residual charge still has its effect upon penetration, as does, in addition, the particle size resulting from hydration.

The fact that uncharged particles enter a cell more readily than ions do suggests that the plasma membrane is charged. Anions are attracted to positively charged areas and cations to negatively charged ones. Since both anions and cations are affected, the plasma membrane is considered to be *a mosaic* of negative and positive charges. However, the overall charge on the plasma membrane is positive, so that anions would be expected to enter cells more readily than cations. This is found to be the case. (Some cells appear to be freely permeable to anions while at the same time they largely exclude cations.' Other cells may be freely permeable to cations but may largely exclude anions.)

Table 11.4. Penetration and Properties of Some Cations.

Property	*Ion*		
	Li	*Na*	*K*
Atomic weight	6.939	22.997	39.096
Degree of hydration	most	intermediate	least
Effective diameter	largest	intermediate	smallest
Entry into cells	slowest	intermediate	fastest

The passage of cations and anions from an external medium into a cell is thought to be in exchange for an ion of like charge inside the cell. Hydrogen ions resulting from metabolic activity are always available for exchange with cations (cation exchange diffusion), and bicarbonate ions of similar origin are available for exchange with anions (anion exchange diffusion). For maintenance of the electric potential across the plasma membrane, equal numbers of anions and cations need not be taken up by the cell, but the exchange of ion for ion must be equal.

All of these considerations on the permeability of cells to ions apply only to situations in which no marked accumulation of some ions occurs and when no active transport mechanism seems to be involved. When marked accumulation occurs, as for example, that of potassium in most cells, an active transport system is suggested.

OSMOTIC RELATIONS IN THE PRESENCE OF PROTEIN

Proteins form ions that penetrate neither natural nor artificial membranes with small pores. When a potassium proteinate solution, at the appropriate pH, is placed in a cellophane bag and immersed in a solution of potassium chloride, the initial distribution of ions is similar to that shown in Figure elsewhere in this chapter. At this time the initial concentration of potassium ions inside is designated as C_i, the concentration of chloride is zero, and the concentration of proteinate is designated as C_i. The initial concentration of potassium ions outside at this time is designated as C_o, and the concentration of chloride ions is designated as C_o.

Since chloride ions penetrate the membrane, they diffuse inward in response to the concentration gradient. To maintain electric neutrality, potassium ions accompany them. At equilibrium, therefore, a certain concentration, X, of potassium and chloride ions will have entered the membrane. At that time the concentration of potassium ions inside the membrane will be $(C_i + X)$; that of the chloride ions inside will be X. The concentration of potassium ions outside the cellophane bag will be $(C_o - X)$ and that of the chloride ions outside the bag will be $(C_o - X)$.

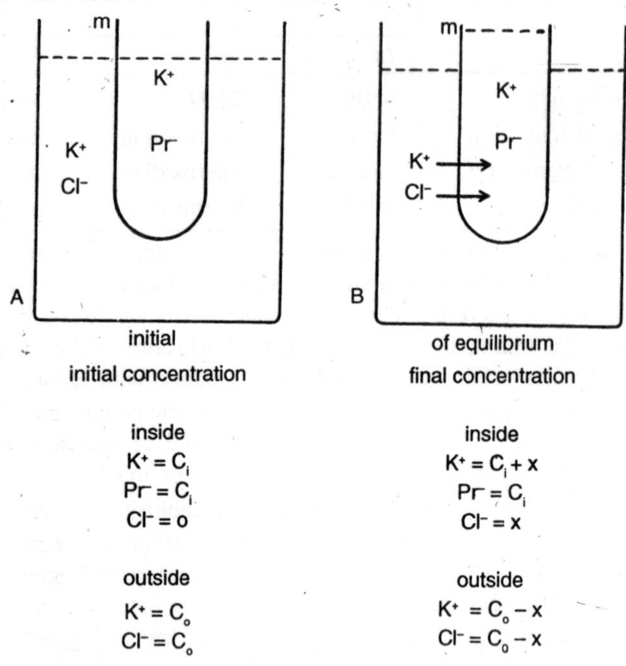

Figure 11.8: Donnan equilibrium. The membrane (m) within which the protein solution is contained is impermeable to protein ions but permits all the other ions to pass.

Donnan, by applying the law of mass action, found that at equilibrium the product of the concentrations of the diffusible ions inside the bag equals the product of the concentrations of the diffusible ions outside the bag:

$$[K^+]_{in}[Cl^-]_{in} = [K^+]_{out}[Cl^-]_{out} \quad (11.1)$$

The physical chemistry and thermodynamics of the Donnan equilibrium are dealt with elsewhere. Substituting for these symbols the concentration terms used before, for the simplest case, in which the compartments are equal, one can write the following equation:

$$(C_i + X)X = (C_o - X)(C_o - X) \quad (11.2)$$

Solving for X:

$$X^2 + C_iX = Co^2 - 2C_oX + X^2 \quad (11.3)$$

$$X = \frac{Co^2}{C_i + 2Co} \tag{11.4}$$

Knowing the initial concentrations of ions inside and outside a membrane, one can use this equation to solve for X, the concentration (at equilibrium) of the diffusible ions which have entered the membrane. The equilibrium is called the *Donnan equilibrium* in honor of its discoverer.

If a mixture of uni-univalent, diffusible and membrane-penetrating electrolytes such as NaCl, KNO_3, LiBr and HI is used instead of KCl inside and outside the tube containing the protein, the following relations hold:

$$\frac{[Na^+]_i}{[Na^+]_o} = \frac{[K^+]_i}{[K^+]_o} = \frac{[Li^+]_i}{[Li^+]_o} = \frac{[H^+]_i}{[H^+]_o} =$$

$$\frac{[Cl^-]_o}{[Cl^-]_i} = \frac{[NO_3^-]_o}{[NO_3^-]_i} = \frac{[Br^-]_o}{[Br^-]_i} = \frac{[I^-]_o}{[I^-]_i}$$

THE EFFECTS OF TEMPERATURE OF PERMEABILITY—ENERGY OF ACTIVATION

If certain substances enter a cell in response to a concentration gradient between the outside and inside of the cell, and if the molecules are freely able to enter and leave the plasma membrane, the temperature coefficient for entry of these substances into cells should be like that found for free diffusion, namely, 1.3 to 1.4.

Experiments show this to be true in a few cases. For example, a temperature coefficient of this order is found for the entry of propyl alcohol into erythrocytes and of some aldehydes into luminous bacteria, suggesting that these molecules move freely through these cell membranes.

However, in many other cases, e.g., the entry of dyes into *Arbacia* or into frog skin, the rate of entry of compounds into a cell is increased several fold by a 10°C rise in temperature, indicating that most molecules have to be activated to cross some barrier in the plasma membrane.

Assuming that the plasma membrane is a sandwich of two layers of lipid between an outer and an inner layer of protein, as was previously discussed, the following illustration will perhaps clarify some of the problems involved.

The plasma membrane of the cell presents the main hurdles to any entering molecule which must pass from the aqueous external medium into the cell membrane, through the plasma membrane and, finally, from the plasma membrane into the aqueous medium of the cytoplasm within the cell. The plasma membrane, as was shown, contains a considerable amount of lipoidal material. If a molecule entering a cell has a high lipoid solubility (that is, a high partition coefficient), neither its entry into nor its transport through the membrane presents an obstacle. However, entry of the molecule from the membrane into the aqueous cytoplasm does present an obstacle because the cohesive bonds it has formed with the lipoid molecules of the membrane must be broken before it can leave the membrane.

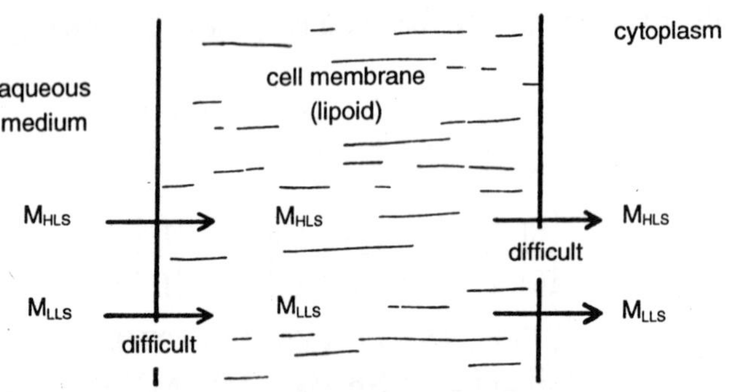

Figure 11.9: Hurdles encountered by a molecule with high lipoid solubility (M_{HLS}) and a molecule with low lipoid solubility (M_{LLS}) when passing through the cell membrane to the cytoplasm.

If, on the other hand, a molecule entering a cell membrane is highly soluble in water (that is, it has a low partition coefficient), its strong hydrogen bonds with the water in which it is dissolved outside the cell must be broken before the molecule can enter the lipoid membrane. Its movement through the lipoid membrane also presents an obstacle. However, such a molecule passes readily from the membrane into the aqueous cytoplasm inside the cell where it forms new hydrogen bonds. In this case, therefore, passage into and through the cell membrane is the most difficult hurdle.

If the energy required to get across any of these hurdles is large, only the molecules with the required energy of activation are successful in moving through the membrane. For this reason, temperature coefficients for permeability may be quite large even when active transport is not involved.

EFFECT OF RADIATION OF PERMEABILITY

(The effects of visible light, ultraviolet radiations and ionizing radiations upon the penetration of substances into cells has been studied, and some literature on this topic exists Some of the findings are controversial, and therefore they will not be reviewed here, but it is known that when cells are sufficiently injured by radiations their permeability to various solutes increases as it does following any other injury.

THE EFFECTS OF SALTS ON PERMEABILITY—SALT ANTAGONISM

The salt environment to which a cell is subjected affects the permeability of the plasma membrane to other substances To study these effects, the electrical conductivity of cells has been measured, as an index of entry of ions. With a cork borer, discs of tissue cells of the kelp *Laminaria* are cut from the blade of the plant. These are stacked like a roll of coins between two electrodes, and the electrical resistance (the reciprocal of the conductivity) is determined.

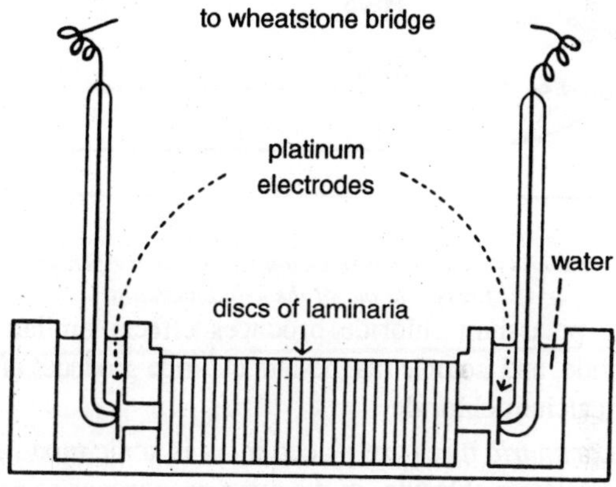

Figure 11.10: Apparatus for conductivity studies.

A decrease in resistance means that the ions are passing through the cells more readily and that the permeability has therefore been increased. A few samples of the many interesting studies performed are cited in the following paragraphs. When *Laminaria* discs are immersed in sodium chloride solutions isotonic with sea water (0.52 molar), their electrical resistance gradually decreases, indicating an increase in permeability.)

After extended exposure the cells, having become injured, are somewhat freely permeable to sodium chloride, suggesting that the plasma membrane of the cells now presents little resistance to entry and exit of the sodium and chloride ions. If not subjected to too long an exposure to the sodium chloride, the *Laminaria* discs recover when immersed again in sea water, and their electrical resistance increases to some extent.

When this experiment is repeated with calcium chloride isotonic to sea water (0.278 molar solution), the electrical resistance of the discs first increases, indicating decreased permeability to the salt, and then falls, showing that the cells, having become injured are now more permeable to the ions.

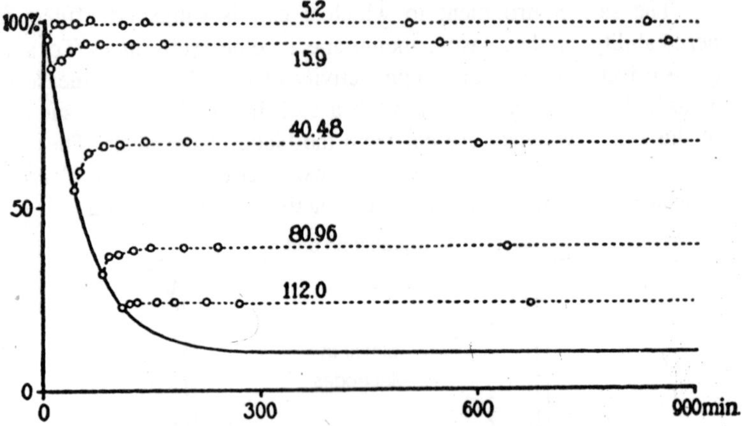

Figure 11.11: The effects of isotonic NaCl on the permeability of disks cut out of the kelp Laminaria.

Isotonic potassium chloride produces effects similar to those of sodium chloride, and isotonic magnesium chloride produces effects similar to those of calcium chloride.

When *Laminaria* discs are placed in an isotonic mixture of sodium chloride and calcium chloride, in the same proportions as present in sea water, the electrical resistance remains about the same as that of the control in sea water (17 hours, in one test), indicating that the permeability is practically normal and that the cells have not been injured.

Since calcium and sodium chlorides have opposite initial effects on the permeability of the plasma membrane, calcium chloride decreasing it and sodium chloride increasing it, these two ions are said to antagonize each other. The permeability of the membrane is completely normal if potassium and magnesium chlorides are also added to the solution of sodium and calcium chlorides in the proportions in which these cations are found in sea water. Salt antagonism appears to be the antagonistic action of different salts in maintaining normal permeability of the plasma membrane.

Similar salt antagonism is also observed on animal cells. For

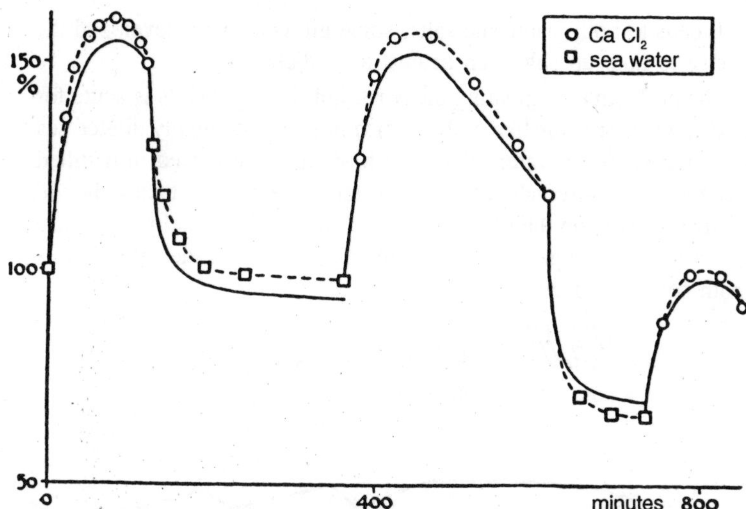

Figure 11.12: The effect of isotonic CaCl2 on the permeability of disks cut out of the kelp Laminaria. The curves show the net electric resistance (ordinate) in per cent of the initial resistance of Laminaria agardhii in 0.278 molar $CaCl_2$ and in sea water. The unbroken line represents calculated values, and the broken line represents observed values. The curves are based on an average of 10 or more experiments.

instance, a frog heart placed in pure sodium chloride solution ceases beating, but if calcium chloride is added in the proportion found in blood (or sea water), the heart revives. However, the beat is not normal unless potassium chloride and a buffer are added also.

Ordinarily glucose is also added to the solution to supply nutrient to the heart muscle cells. Such a balanced salt solution was developed for frog heart by Sidney Ringer, a British physician, and is called *Ringer's solution* in his honor. However, it is not known whether the salts exert their action by affecting the permeability of heart muscle cells or for other reasons; e.g., calcium may be required in the bathing medium because of its inciting action on muscle contraction. This topic is discussed in other chapter of this book. Many similar solutions have been developed since for experimental use with protozoans, marine eggs, blood cells, tissues, organs and tissue cultures.

The fact that calcium decreases the pe eability of the plasma membrane to water and to other substances is of great importance, especially in the water relations of cells of fresh water organisms. Such organisms must maintain within them a certain amount of salt for normal functioning, even though little salt is available in the bathing medium and even though such cells continually take up water and lose salts. Experiments show that calcium ions in water reduce the permeability of gill cells to both water and salts; some gill cells have developed the capacity to take up salt even against a gradient.

The problem of regulating the permeability of gill cells is acute for brackish water animals like crabs, which migrate into highly diluted sea water or even fresh water. When such animals are placed in distilled water they may lose salts and gain water so rapidly that they die. In hard fresh water, on the other hand, they are able to get along because calcium reduces the permeability of the plasma membranes of gill cells to both solutes and water.

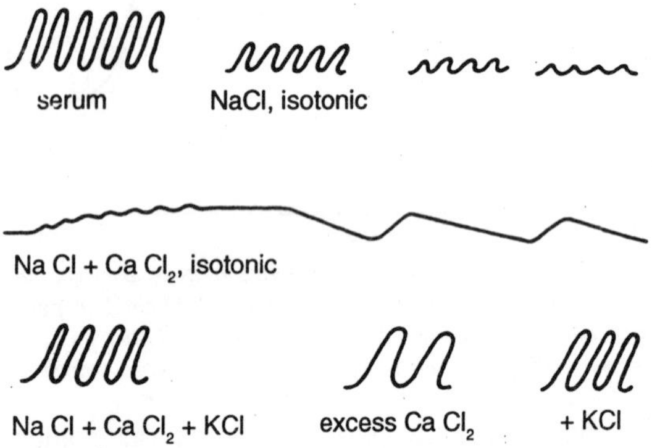

Figure 11.13: The effects of salts on a frog's heart. The heart stops in isotonic NaCl, but beats slowly when $CaCl_2$ is added in the proportions found in blood (the heart must be stimulated once). The beat is almost normal when KCl is added in the proportion found in blood and when the mixture contains a buffer. Excess $CaCl_2$ prolongs the beat of the heart and is antagonised by KCl.

The antagonistic relations between various cations are more complex than would be suggested by these discussions. They have been extensively studied, but a satisfactory interpretation in terms of molecular interactions in the cell membrane has not been found. Although anions of salts, used in balanced media, probably also affect the life of the cells suspended in them, salt antagonism is thought to be essentially a cationic phenomenon.

THE EFFECTS OF NARCOTICS AND ANESTHETICS UPON PERMEABILITY

Narcotics and anesthetics generally seem to exert their characteristic effects on cell sensitivity and respiration in proportion to their lipoid solubility, indicating that they may need to enter the cell to affect it. Some values for the narcotic action of a number of compounds, correlated with their solubility in olive oil, are given in Table elsewhere in this chapter.

Anesthetics also affect the permeability of cells in an interesting manner. For example, butyl alcohol decreases the penetration of glycerol into the red blood cells of human beings, rats, rabbits, guinea pigs and birds, but it increases the penetration of glycerol into the red blood cells of sheep, horses, dogs, bats, pigs, reptiles and fishes.

Thiourea, on the other hand, has effects opposite to that of butyl alcohol on the penetration of glycerol into the red blood cells of the two series of animals tested. Furthermore, the same anesthetic, at different concentrations, may have the reverse effects on the permeability in a cell. For example, urethan in low concentration decreases the permeability of discs of algal tissue, whereas at high concentrations it has the opposite effect when first applied. In both instances prolonged exposure results in injury and increased permeability.

Table 11.5: Relation Between Narcotic Action and Lipoid Solubility

Compound	*Vol. % Required to Just immobilise Mice*	*Solubility in Olive Oil at 37° C*
Ethane	80	1.3
Acetylene	65	1.8
Dimethyl ether	12	11.6
Methylchloride	6.5	14.0
Dimethyl acetal	1.9	100
Chloroform	0.5	265

Such diverse effects of anesthetics on the permeability of cells are difficult to explain. Perhaps the anesthetic molecules, being highly lipid soluble, accumulate at the surface of the cell, thus altering the properties of the membrane by either increasing or decreasing the ease of entry of other molecules. For example, if the solution of anesthetic in the membrane makes the membrane more lipoidal in nature, molecules of high partition coefficient will enter more readily than before, whereas those of low partition coefficients will enter less readily. The converse would also be possible.

THE EFFECT OF THE PHYSIOLOGICAL STATE

Movement of solutes through the plasma membrane is dependent primarily on the physiological state of the cells. Active cells utilize cell foods and accessory materials, thus maintaining a continual diffusion gradient for these substances, enabling the substances to enter the cell readily. When a muscle cell or nerve cell is shocked and stimulated, its permeability to ions is increased.

An active muscle cell is permeable to amino acids, glucose and other materials, whereas a resting one is not. Electric stimulation of various plant and animal cells results in an increase in their permeability to solutes at the negative electrode, as shown by increased conductivity (decreased resistance). The leakage of potassium ions at this electrode in stimulated *Nitella* and the squid giant axon may be sufficient even for measurement by microchemical methods.

As shown by experiments with radioactive sodium, sodium ions also pass through the membrane of a stimulated cell more freely than they do through the membrane of a resting cell. Injury of the cell by heat, radiations, overstimulation, anesthetics, pH change, salt unbalance or any other means results in an increased permeability of the cell By use of the conductivity method Osterhout has made an extensive study of the effects various degrees of injury have on permeability of *Laminaria* and other algae.

If the injury is slight there is only a slight increase in the conductivity, indicating that the permeability of the cells has increased slightly. If the injury is large, the increase in conductivity is greater. If exposure to any of the injurious factors just listed has not been excessive before the cells are returned to normal conditions, complete re very of normal permeability often follows)

However, as shown in Figures elsewhere in this chapter, only partial recovery may occur, and the permeability may be permanently increased by injury. All gradations of change are possible, depending upon the damage to the cells. Studies of permeability changes following injury of a variety of cells indicate a similar action. Death, as measured by permeability, is a quantitative phenomenon.

FACILITATED DIFFUSION

When the rate of entry of a solute into a cell, e.g., galactose into some red blood cells, increases proportionally with increase in solute concentration in the medium only to a limit, beyond which there is little further change in rate of entry with further increase in concentration, permeation of the solute is considered to be limited by saturation of a *carrier system*. The carrier is perhaps some substance in the plasma membrane which combines with the solute in question, transporting it into the cell. Presumably the carrier liberates the solute inside the cell and itself returns to "*pick up*" another molecule (or ion). This type of transport is often called *facilitated diffusion.* Facilitated diffusion occurs at the expense of kinetic energy alone; only those solute molecules with the critical increment of kinetic energy are able to penetrate through the plasma membrane. Facilitated diffusion is to be distinguished from active transport in which a carrier may also be involved, but passage of the solute through the plasma membrane occurs at the expense of metabolic energy.

Facilitated diffusion is therefore not affected by metabolic poisons, whereas active transport is. Facilitated diffusion may show a high temperature coefficient. This is to be expected because passage of the solute through the plasma membrane is dependent upon the number of solute carriercomplexes having sufficient energy for the purpose.

The proportion of complexes having the critical energy may increase several fold with an increase in temperature of 10° C. Although facilitated diffusion does not alter the final equilibrium which could be attained even in the absence of a carrier system, it does greatly speed the attainment of the equilibrium. It is, therefore, advantageous to the cell.

EXCHANGE DIFFUSION

When the plasma membrane of a cell is permeable to given molecules or ions it is conceivable that some will come out of the cell at the same time as some go in, especially if a carrier mechanism transports the molecules (or ions). Thus, when some cells are kept in a solution of ^{24}Na for a period, then placed in "cold" sodium ion solution, the radioactive sodium leaves the cell in proportion to its relative concentration in the cells.

The loss of ^{24}Na presumably occurs by exchange diffusion, a radioactive sodium ion being exchanged at random for a cold sodium ion, because if the bathing solution containing sodium is replaced by one of isotonic mannitol, the exchange ceases since no sodium ion is available for exchange. The exchange is at the expense of kinetic energy, not metabolic energy, as shown by its independence of metabolic poisons.

Exchanges of one molecular (or ionic) species for another are also known to occur when the two species are transported by a common carrier. For example, the amino acids histidine and tryptophan are readily exchanged for each other by Ehrlich ascites tumor cells. These exchanges do not require metabolic energy nor are they subject to the inhibitors which affect metabolically linked transport. Some cation-for-cation and anion-foranion exchanges require metabolic energy.

COMPARISON OF THE PERMEABILITY OF DIFFERENT KINDS OF CELLS

The relative permeability of different kinds of cells to a particular substance may be compared by any method which is equally applicable to all of the cells. For example, the rate of hemolysis has been used for comparison of the permeability of different erythrocytes. Erythrocytes of various species of mammals, when tested in this manner for their permeability to glycerol, show differences covering a range of 240 fold. Similar differences are observed for entry of sodium ions, potassium ions, urea and glucose.

When comparing the permeability of various kinds of cells to a given substance, and the same method cannot be applied to all of the cells, then it is necessary to measure the amount of the substance entering a unit area per unit of time and per unit concentration difference between the inside and outside of the cells.

Table 11.6: Variation in Permeability of Erythrocytes of Various Mammals, Measured by Time for 75 Per Cent Hemolysis in 0.3 Molar Glycerol in 0.12 Per Cent NaCl

Animal	*Time*	*Animal*	*Time*
Rat	3.5	dog	253
Man	5.1	pig	340
Mouse	12.9	cat	459
Rabbit	21.8	ox	612
Guinea pig	38.2	sheep	850

Calculation of a permeability constant incorporating these quantitative data has been made possible by means of certain equations. These are based on Fick's law of diffusion, which states that a substance, S, will diffuse through an area, A, at a rate, dS/dt (amount, dS, per unit time, dt), which is dependent upon the difference in concentration between the substances at a certain distance apart.

$$\frac{dS}{dt} = DA\frac{dC}{dx} \tag{11.6}$$

In this equation D is the diffusion coefficient and is given in moles per unit area per unit concentration gradient, and dC is the difference in concentrations ($C_1 - C_2$) over the distance dx.

For passage through a membrane such as that of a cell, equation 11.6 becomes

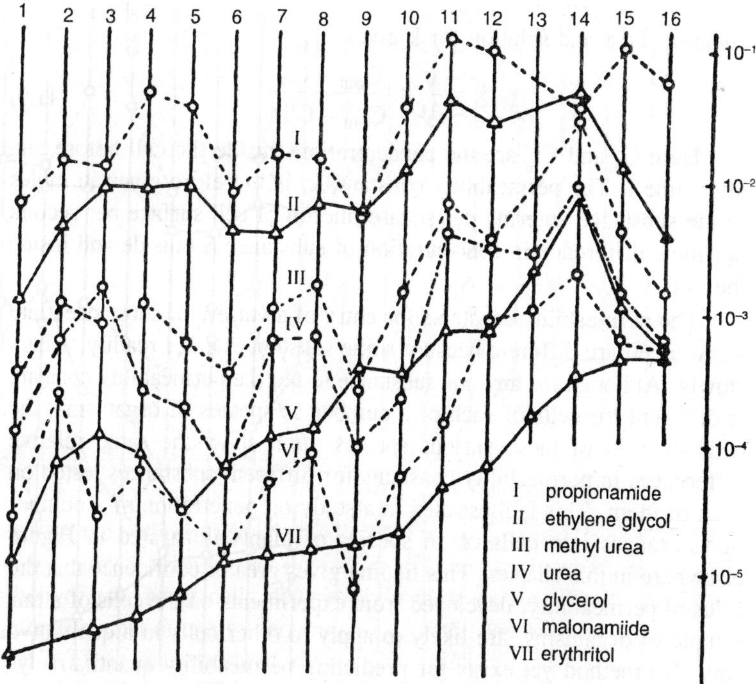

Figure 11.14: Permeability constants of plant cells to nonelectrolytes. Cells from 16 species are represented: 1, Leaf cells of Plagiothecum denticulatum; 2, Oedogonium sp.; 3, root cells of Lemna minor; 4, Pylaiella litoralis; 5, Zygnema cyanosporum; 6, subepidermal cells of Curcuma rubricaulis; 7, Spirogyra sp.; 8, leaf cells of Elodea densa; 9, epidermal cells of Rhoeo discolour; 10, epidermal cells of Taraxacum pectinatiforme; 11, "leaf cells" of Chara ceratophylla; 12, internodal cells of Ceramium diaphanum; 13, Bacterium paracoli; 14, Oscillatoria princeps; 15, Melosira sp.; 16, Licmophora sp.

$$\frac{dS}{dt} = \frac{DA}{M}(C_{out} - C_{in}) \qquad (11.7)$$

Here C_{out} represents the concentration of the substance, S, outside, and C_{in} is its concentration inside the cell. A is the area of the membrane, and M-which is substituted for dx in equation 11.6-represents the thickness of the membrane.

Since the amount of substance (S) per volume (V) of the cell is Ci_n, $C_{in}V$ may be substituted for S. When k, the permeability constant, is substituted for D/M, the equation is written as follows:

$$\frac{dC_{in}}{dt} = k\frac{A}{V}(C_{out} - C_{in}) \qquad (11.8)$$

If the volume of fluid outside the cell is large, integration of equation 11.8 and solution for k gives

$$k = \frac{V}{At} \ln \frac{C_{out} - C_{in}}{C_{out} - C^{1}in} \qquad (11.9)$$

Here C_{in} and C^{l}_{in} are the concentrations inside the cell before and after time t. The permeability constant, k, is therefore given in moles of the substance entering per square micron of cell surface per second per mole difference in concentration of substance S outside and inside the cell.

The permeability constants for entry of a variety of substances into a given cell are different because some substances enter readily, others slowly. Although for any one substance tested the permeability constant is different for cells of each of a number of species of organisms, the cells of each of these various species often show the same relative differences in permeability constants for different substances tested on each of them. This is illustrated in a study on penetration of a number of substances into cells of 16 species of plants illustrated in Figure elsewhere in this chapter. This finding gives greater confidence that the rules of permeability, developed from experiments on the cells of a fair sample of organisms, are likely to apply to other cells in a qualitative way. No method yet exists for predicting permeability quantitatively.

Intracellular Transport

Transport also occurs between organelles in a cell. Such transport may occur by the same mechanisms as between the cell and its bathing medium, although information is at present rather limited.

METABOLICALLY LINKED TRANSPORT THROUGH CELL MEMBRANES AND PINOCYTOSIS

Active Transport

When a substance moves through a cell membrane against a concentration gradient, either into or out of a cell, it is at once apparent that such movement could not be by diffusion powered by kinetic energy alone. To move the molecules of a substance against a concentration gradient and (or) against an electrochemical gradient requires a mechanism other than-or in addition to-diffusion, and a source of energy for such movement.

The cell must do work in moving the substance through the cell membrane against a concentration gradient. Any movement of substances through the cell membrane which requires work on the part of the cell is called *active transport* or more accurately, *metabolically-linked transport*.

The phenomenon of active transport is perhaps best illustrated by the activity of the cells of the kidney tubule. Pieces of kidney tubule, separated with a micromanipulator from a chick embryo, often round up into spheres one cell layer thick. When such spheres are immersed in dilute solutions of phenol red, dissolved in embryonic tissue fluid, each sphere accumulates the dye in its internal cavity so that the cavity becomes deep pink in colour.

The dye may be seen passing through the cells of the tubule in which it lingers in transit and where it becomes orange in colour, a change to be expected of phenol red at the pH (about 6.7 to 6.8) inside the cells. If phenol red were diffusing through the cells in response to the difference between the concentrations of the dye in the bathing solution and in the vesicle, it would move inward until equilibrium was established.

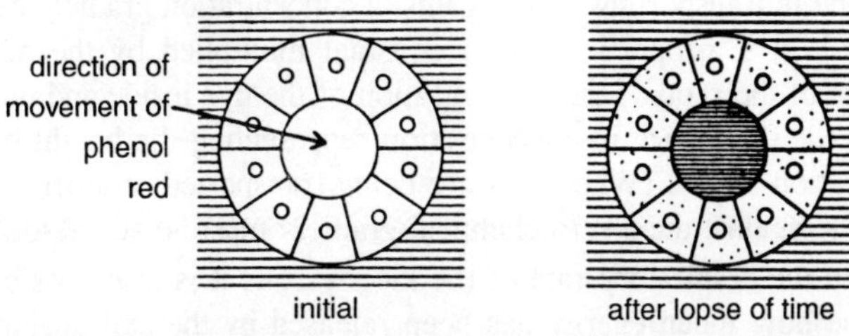

Figure 11.15: Accumulation of phenol red in sections of proximal kidney tubules of embryo chick (diagrammatic). Phenol red moves inward until it becomes more concentrated in the inside of the vesicle than on the outside.

The appearance of the dye in much higher concentration inside the cavity of the vesicle than elsewhere means that the dye is passing through the cells and is being extruded or secreted into the cavity against a concentration gradient.

If the kidney cell preparation is placed on ice at the time phenol red is added, it fails to accumulate the dye. If it is placed on ice after accumulation of the dye inside the vesicle, the dye then diffuses out and distributes itself equally between the medium and the lumen of the vesicle.

Warmed up to chick body temperature (39°C.), the vesicle again accumulates the dye. The experiment demonstrates that although the dye continues to diffuse through tubule cells even at low temperature, it is no longer accumulated, probably because some process essential to accumulation is inhibited. Measurements show that the respiration of the kidney tubule cells comes virtually to a halt when the tubule is placed on ice. Since respiration is the main source of the energy by which cellular work is accomplished, the inference is drawn that the movement of the dye against a concentration gradient requires cellular work.

This inference is corroborated by experiments, again using dyes, in which respiration of the kidney tubule cells, even though at 39°C., is inhibited by a metabolic poison. When respiration virtually ceases, dye is no longer accumulated. Diffusion, however, continues in response to the concentration gradient, and a preparation which is poisoned after it has already accumulated dye is still able to lose it to the medium. These experiments demonstrate that kidney tubule cells can do work against a concentration gradient only if they are able to liberate energy.

Energy, then, is needed to carry molecules of dye against a concentration gradient, just as energy is needed to climb a hill. The work done in transporting a dye against a concentration gradient can be calculated. It is proportional to a constant multiplied by the natural logarithm of the ratio of the concentration of the dye inside and outside the cell, the difference in concentration representing the height of the concentration "*hill*". (When ions are being transported, a term for the work done against the electrochemical gradient must be added to this.)

From the respiratory rate of the kidney cells it is also possible to determine how much energy has been released by the cell during the time that a given amount of solute has been transported. The calculations show that the energy released in respiration is more than enough to do this work. Two criteria have often been used to determine whether a substance moves across the plasma membrane by active transport: first, whether it moves against a concentration gradient; second, whether it requires metabolic energy for the movement. If a substance moves against a concentration gradient, transport is probably active, although not all active transport need be against a concentration gradient.

However, it is usually necessary to demonstrate that the substance in question is osmotically active inside the cell and that its movement really occurs against a concentration gradient. That some substances are present inside cells in a concentration greater than outside does not necessarily imply that they enter the cell by an active process against a concentration gradient. For instance, a substance may be rendered osmotically inactive inside the cell in any of a number of ways. If ionic, it might unite with another ion to form an insoluble compound (e.g., calcium oxalate). Or it might possibly be dissolved in lipid and therefore

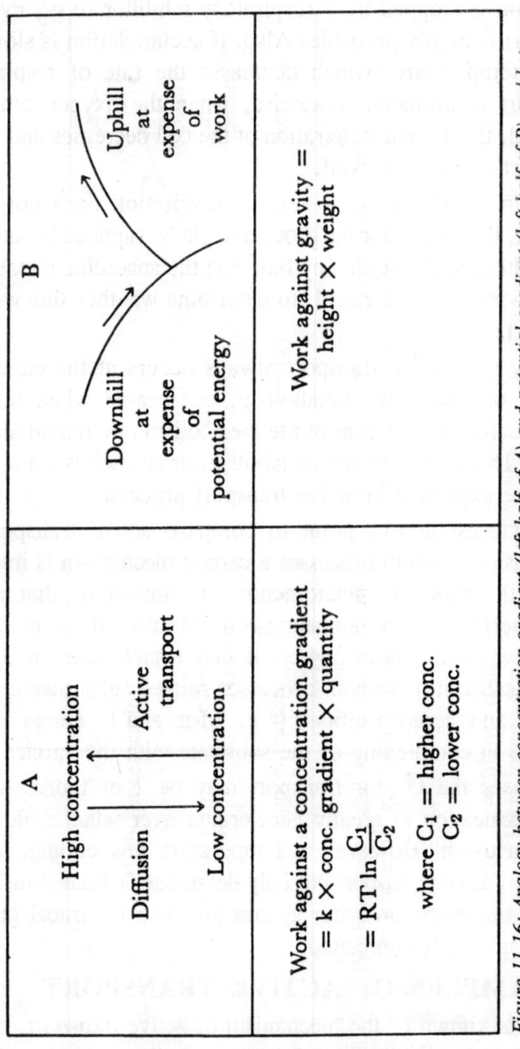

Figure 11.16: Analogy between concentration gradient (left half of A) and potential gradient (left half of B) and between movement against a concentration gradient (right half of A) and work done in moving uphill (right half of B). Equations for work against a concentration gradient (osmotic work) and for work in lifting a weight up a height are given at the bottom of the figure.

be unavailable for osmotic action in the aqueous continuum of the cell. Organic substances like glucose may be polymerised to form insoluble glycogen or starch, and glucose molecules could thus be trapped, that is, taken out of solution, and a concentration gradient favourable for uptake might be maintained in spite of extensive entry.

Since active transport depends on release of energy from metabolic reactions, the effect of respiratory activity on uptake of the materials by a cell should be tested. If the accumulation of a substance ceases after respiration is stopped by a respiratory inhibitor (e.g., cyanide), active transport is highly probable. Also, if accumulation is slowed by lowering the temperature (which decreases the rate of respiration), active transport is probable. Likewise, when the oxygen tension is greatly reduced, the aerobic respiration of the cell decreases and usually active transport declines as well.

In cases in which blocking aerobic respiration does not inhibit active transport, the energy for the process might be supplied by anaerobic metabolism. This could be tested by blocking the anaerobic metabolism, for example, with iodoacetic acid, to determine whether this will stop active transport.

It is likely that active transport always occurs at the expense of either aerobic or anaerobic metabolism, but ceases when both are stopped. As much as 30 per cent of the metabolism of a mammalian red blood cell and 15 per cent of the metabolism of an ascites (tumor) cell is thought to be expended in active transport processes.

It is of interest at this point to compare active transport and facilitated diffusion. In both processes a carrier mechanism is involved. Therefore, both show the phenomenon of saturation; that is, the permeability increases with increasing concentration of the compound until the carrier is translocating all it can (much like an enzyme saturated with substrate). In both processes related substrates compete for the carrier, and stereospecificity (e.g., of n- and L- compounds) is of consequence in complexing of the substrate with the carrier.

In both cases the Q_{10} for transport may be 2 or more. In both cases, also, permeation is greatly accelerated over what could occur by ordinary diffusion. However, a temperature low enough to stop metabolism stops active transport but only decreases facilitated diffusion. Finally, active transport can occur against an electrochemical gradient whereas facilitated diffusion cannot.

EXAMPLES OF ACTIVE TRANSPORT

Not much is known of the mechanism of active transport, but the process seems to be quite general in plant cells, animal cells and microorganisms. It is thought that probably most solutes enter cells actively. A number of interesting cases are cited here.

Accumulation of Potassium in Cells

A great accumulation of potassium is found in many plant and animal cells and in microorganisms. For example, *Nitella, a* fresh water plant with large cells, has 1065 times as much potassium as the water in which it grows; *Valonia,* another large plant cell, has 41.6 times as much; and *Chara* has 63 times as much. Although the concentration of potassium is not always this much greater inside than outside the cell, such concentrations are characteristic of plant cells generally. A somewhat similar difference in concentration of potassium between the inside of the cell and the bathing medium is found in muscle and nerve cells and in microorganisms such as yeast.

That potassium is present in cells as an ion and is osmotically active is demonstrated by measurement of the electrical conductivity. The electrical conductivity of cells is of such a magnitude as to indicate that potassium exists in the cells in dissolved, ionic form; otherwise, the conductivity would be much lower since potassium is the main cation of the cell.

The potassium gradient between the inside of the cells and the medium therefore could be maintained only by active transport. If this is the case, potassium should leak out of the cells, in response to the concentration gradient, when metabolism is inhibited. This occurs; for example, if either a squid giant axon or a *Halicystis* cell is placed under anaerobic conditions, potassium leaks out because the energy liberated by aerobic metabolism is no longer available to keep it inside.

If oxygen is bubbled through the water, it enables the cells to take up potassium again. In these cells anaerobic metabolism apparently does not provide the energy required for active uptake of potassium. Lowering the temperature to near the freezing point permits potassium to escape from and sodium to enter human erythrocytes, presumably because the metabolic reactions required for potassium uptake are suspended.

When the temperature is raised, potassium is again taken up. On the other hand, the human erythrocyte continues to hold its potassium even when the oxygen tension is greatly decreased at room or body temperature. This implies that in this case an anaerobic process (called glycolysis in animal cells) liberates the energy for active transport of potassium.

Such a postulate is corroborated by the finding that poisoning the anaerobic glycolysis with fluoride causes potassium to leak, out of the cells, even at body temperature. Yeast holds its potassium under anaerobic conditions, provided glucose is present. It is unable to liberate energy anaerobically (by fermentation) unless nutrient is present in the bathing medium; i.e., it does not use its stored food for this purpose.

Therefore, when yeast is washed free of glucose, and kept in the absence of oxygen, it loses its potassium. When glucose is restored, potassium also is recovered. A similar phenomenon is observed for *E. coli*. A normal, healthy cell membrane is necessary for maintenance of the difference in concentrations of potassium between the inside and the outside of the cell.

Alteration or injury of the membrane or of the cell leads to potassium leakage. For example, when the giant axon of the squid is stimulated, potassium leaks out. The same is true of *Nitella*. In other words, when the membrane is temporarily altered, ions can get through. On recovery, however, potassium is reabsorbed. Potassium leakage following stimulation of cells has been the subject of a number of studies.

However, it has been found that potassium is not the only ion which moves through the membrane of a stimulated cell. By carefully timing the movements of radioactive sodium and potassium in separate experiments one can observe that in a stimulated cell sodium moves in, then potassium leaks out. The sodium subsequently leaves and the potassium once again enters.

Analysis of this occurrence suggested, as a possible explanation, that potassium is not accumulated by active transport, but rather that sodium is extruded by active transport; this process is referred to as the *sodium pump*. Elimination of sodium in this manner presumably goes on in all cells which accumulate potassium, and supposedly the elimination of the sodium creates a cation gradient favourable for the passive uptake of potassium.

In such a case, when potassium leaks out of cells which have been inhibited by metabolic poisons, low temperature or anaerobiosis, it could be considered that the sodium pump is temporarily out of action and sodium enters the cells by diffusion. Sodium is thought to be extruded only when metabolic energy for this purpose is liberated by the cell (in the absence of poisons and at appropriate temperatures). However, evidence indicates that the sodium pump explanation alone is not in keeping with the facts.

Thus, not only is sodium actively extruded in many cases, but potassium is actively transported into the cell as well, since sodium extrusion is linked to potassium uptake. For example, in the red blood cell the electrochemical gradient is inadequate to bring in the potassium, and active transport is physically necessary. It would also appear likely, in some cases, that one mechanism handles both sodium and potassium because changing the concentration of one of the ions outside the cell also affects the transport of the other ion. Evidence from use of tracers indicates that potassium is actively transported in muscle and nerve cells as well, even though the electrochemical gradient is favourable for passive entry. Somewhat similar data are found for the cells of the red alga *Porphyra perforata*.

In yeast, potassium is taken up by active transport in exchange for hydrogen ions. It is probable that the ion transport is not the same in cells of different species; for example, the erythrocytes of a few species of mammals, such as the dog and cat, present an exception, accumulating sodium instead of potassium. Nor is ion transport the same even in cells of different tissues in the same species. For example, the epithelial cells of the small intestines transport actively both sodium and chloride ions whereas cells of the kidney tubules transport sodium actively and chloride moves passively in response to concentration gradients.

Root Pressure and Guttation

Considerable information has accumulated on the development of root pressure in plants. Experimental analyses of the sap indicate that the root hair cells are capable of taking up, against a concentration gradient, the various necessary salt ions such as potassium, magnesium, calcium and nitrate. From the root hair cells the salts are transferred to the cortical cells and from the cortical cells they are actively secreted against a concentration gradient into the xylem ducts.

From there the salts may readily be distributed by diffusion to any cells where they are required in the plant. Accumulation of salts in the root cells, thus leading to a high osmotic pressure, accounts for bleeding of a cut stem and for guttation, the phenomenon in which, under humid conditions, droplets of water appear from water-secreting glands along the edge of a leaf. Although a detailed discussion of bleeding and guttation is out of place here, such phenomena indicate that the plant cells actively transport some materials against concentration gradients.

Vacuolar Function

It is possible that fresh water protozoans make use of active transport in voiding water. The water surrounding them contains less salt than the protozoan protoplasm does; therefore, water enters in response to the concentration gradient but must leave the protozoans against the concentration gradient. The magnitude of the elimination of water becomes evident when it is realised that every 15 or 20 minutes a *Paramecium* empties through its contractile vacuole a quantity of water equal to its entire volume. On the other hand, a brackish water protozoan in slightly diluted sea water may take in little water, in which case the activity of its contractile vacuole may be very low.

However, if the medium is greatly diluted, as after a heavy rain, the rate of vacuolar contraction may be greatly increased. When a fresh water protozoan is poisoned with cyanide, and its respiration thus suspended, the contractile vacuoles fail to function and the cell swells. If it is not washed free of cyanide, it bursts.

Active transport is likely in these examples, but to determine the mechanism it was necessary to determine whether, in the formation of the vacuole, salts are first secreted against a concentration gradient so that then water enters passively into the vacuole, or whether water is actively passed into the vacuole directly against a concentration gradient.

Evidence from micropuncture and analysis of the fluid withdrawn from the contractile vacuole of *Amoeba* at various times in the vacuolar cycle suggests that the fluid secreted into the contractile vacuole is isosmotic with the cytoplasm. The salt is subsequently actively absorbed against a concentration gradient, leaving the fluid hypoosmotic to the cytoplasm.

Active Transport in Cells of the Gut

When there is a high concentration of sugars in the gut of a dog, a concentration gradient exists from the cells of the lumen of the gut to the blood stream favourable to movement of sugar by simple diffusion through the plasma membranes. To test whether such movement does occur, the metabolism of the gut cells must be inhibited by appropriate metabolic poisons, so that active transport can be stopped.

When this is done in isolated loops of the intestine, it was found that pentoses move as freely as they did before, but though some glucose and galactose still move into the blood, they do so at a much lower rate than in the absence of poison. This experiment suggests that the movement of glucose and of galactose depends in part upon active transport. To test this hypothesis, the blood stream is loaded with glucose (or galactose) so that any movement of glucose from the unpoisoned gut cells to the blood must then occur against a concentration gradient.

The fact that it continues at a relatively rapid rate indicates that active transport is involved. A similar experiment is one in which a concentration of glucose (or galactose) less than that in the blood is placed in the gut so that any movement from gut to blood must occur against a concentration gradient. It still does. The transporting mechanism however, has a limited capacity. For example, if the glucose concentr-

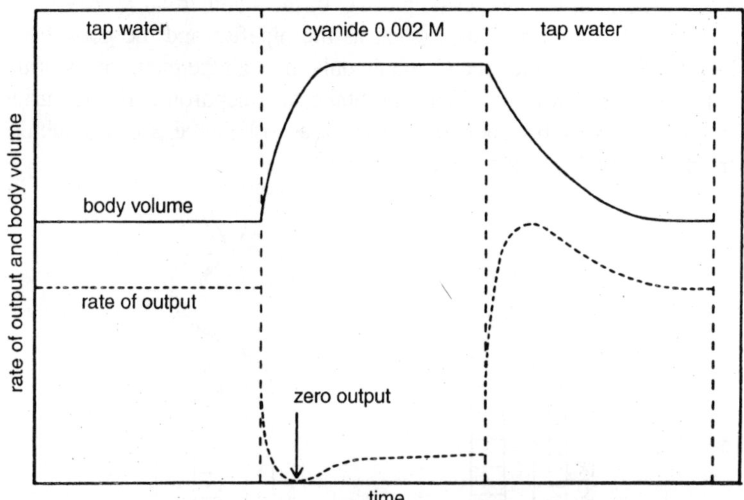

Figure 11.17: The effects of cyanide on the swelling and vacuolar output of the fresh water protozoan Zoothamnioa The time of a complete experiment is about 2 hours.

ation in the gut is increased, the rate of its transport into the blood rises until a definite concentration limit is reached, beyond which further increase in its concentration has no effect on the rate of its transport. The rate of transport is seemingly limited by that amount which the carrier mechanism can transport per unit time (saturation). The transport of sugars requires sodium and appears to be linked to sodium transport.

The conclusion from all of these experiments is that some sugars (e.g., some pentoses), probably move from gut to blood by simple diffusion through the plasma membranes of the gut cells and therefore only when the concentration gradient in that direction is favourable. So also do glucose and galactose to a small extent, but in addition they are actively transported even against a concentration gradient, much of the transport of glucose occurring apparently in this manner.

It is interesting that when two sugars, e.g., D-glucose and n-galactose, which are known to be actively transported, are tested at the same time, they compete with each other. This indicates a limited amount of carrier to which either sugar but not both can be bound at the same time. The binding is stereospecific and neither L-glucose nor L-galactose can be actively transported. Of some 49 sugars tested, only 14 were found to be actively transported in the intestine.

Active Transport in the Cells of the Nephric Organs

Active transport against a concentration gradient is most readily demonstrable in the tubules (nephrons) of the aglomerular kidney found in only a few marine fishes (such as the pipefish and the goosefish), because fluid can enter such a tubule only through the membranes of the cells. The only way in which the tubule of a nephron can form urine in this case is by transport of materials, across the cells of the tubule, from the blood into the lumen.

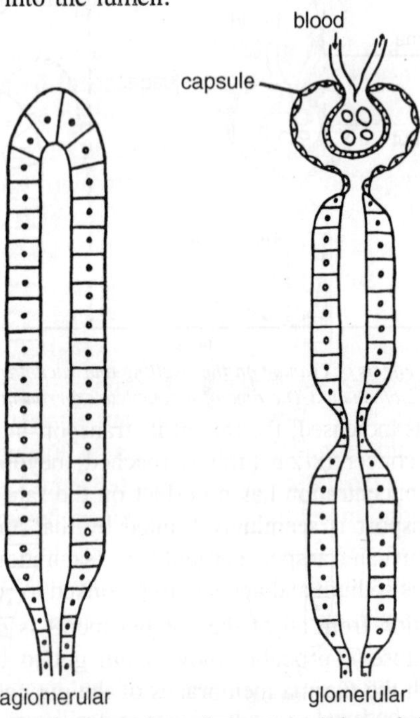

Figure 11.18: Diagram of aglomerular and glomerular kidney tubules (nephrons). Both are copiously supplied with blood capillaries along the outside. Only the glomerular tubule has a capillary tuft in a capsule.

If the concentration of a given compound in the tubule is less than that in the blood, entry of that compound into the tubule can occur by simple diffusion through the cell. On the other hand, the urine of aglomerular fishes is found to contain creatine, creatinine, uric acid, magnesium, potassium, chloride and sulfate, each of them often in a concentration considerably greater than in the blood.

The aglomerular kidney also concentrates in the urine a variety of foreign materials, such as nitrate, thiosulfate, sulfocyanides, indigo carmine, neutral red and phenol red, that have been experimentally injected into the blood. Moreover, this activity ceases when oxygen is withheld or when the metabolism of the kidney cells is inhibited by appropriate poisons. All evidence, therefore, indicates that active transport against a concentration gradient, requires metabolic energy.

The glomerular kidney, which is characteristic of all vertebrates except the fishes cited, contains a glomerulus or tuft of blood capillaries in an epithelial cup for filtration of water and of dissolved constituents of the blood other than the proteins. While the fluid passes along the length of the tubule, part of the water and some of the other filtered constituents are reabsorbed. Although active transport is more difficult to demonstrate in the tubules of glomerular kidneys, it occurs in the reabsorption process (which often takes place against a concentration gradient) as well as in the secretion of certain wastes into the tubules.

Inhibition of the metabolic processes in the cells of the kidney inhibits both of these reactions. Also, some wastes are found concentrated in the urine to a greater extent than would have been possible only by reabsorption of water from the filtrate. These must have been voided into the urine by active transport against the concentration gradient. Decisive and direct proof of such secretion against a gradient has already been shown in sections of the kidney tubules of a chick embryo.

Accumulating evidence indicates that active transport occurs also in the cells of tear glands of marine birds, and in cells of the nephric organs of some invertebrates. For example, in some fresh water invertebrates the urine is found to be hypotonic to the blood. This indicates either the elimination of water against a concentration gradient or the selective uptake of salts against a concentration gradient.

When certain species of crabs (or other invertebrates), capable of tolerating brackish water, are transferred from sea water to diluted sea water, they void more urine, and their respiration increases correspondingly. The increase in each case is larger if the dilution is greater. The increased respiration has sometimes been interpreted as a result of the increased amount of energy required for eliminating the excess water taken up osmotically by the gills, but proof is lacking.

In general, studies on the movement of various materials against concentration gradients in the kidney cells indicate that the materials move at different rates.

The rules governing such movements resemble the rules of permeability previously discussed. Thus, very large molecules are not eliminated (e.g., an aglomerular kidney never passes sucrose, inulin or albumin); movement is related to lipoid solubility, size playing a lesser role; and so on.

ACTIVE TRANSPORT OF ORGANIC NUTRIENTS INTO CELLS

Glucose appears to enter yeast and erythrocytes only when energy is expended, even in the presence of a favourable concentration gradient. However, the evidence for this is indirect and consists in part of the finding that heavy metals, such as copper and uranium, inhibit glucose uptake in these cells.

It is postulated that enzymes at the cell surface are necessary for uptake of glucose, and that these enzymes are selectively poisoned by copper, uranium and other heavy metals. Evidence of a chemically and metabolically coupled transport system for uptake of some substances, such as amino acids, into bacteria has also been presented.

The entry of these substances shows pH dependency, it is poisoned by typical metabolic poisons and the transport is competitive with analogues. Furthermore, the transport system can be saturated. Such a system is feasible for bacteria in which all the enzymes required for aerobic metabolism are located in the surface of the protoplast. In such cases an enzyme would probably transport a molecule by coupling with it at the surface and handing it on to another enzyme in the metabolic main line.

ACTIVE TRANSPORT IN MITOCHONDRIA AND ENDOPLASMIC RETICULUM

Active transport at the expense of metabolic energy occurs not only through the plasma membrane but also through the membranes of the mitochondria, and in some cells (if not generally) in the endoplasmic reticulum and possibly in other organelles. Thus active accumulation of calcium phosphate against a concentration gradient has been observed many times.

As much as a 50-fold increase in calcium of rat kidney mitochondria has been recorded, leading to practically complete exhaustion of calcium in the medium. Such uptake occurs only if an oxidizable substrate is present and stops if the respiration is blocked by a metabolic inhibitor (e.g., cyanide).

Also required is the high energy phosphate compound adenosine triphosphate and inorganic phosphate as well as magnesium ion. Uptake does not occur at 0°C. Inorganic phosphate accompanies calcium ion in being absorbed and is also accumulated. The endoplasmic reticulum of striated muscle cells has also been found an active site of calcium accumulation which obeys all the criteria of active transport. The uptake has been traced autoradiographically to those elements of the muscle cells which are portions of the endoplasmic (sarcoplasmic) reticulum. The rapid and active uptake of calcium removes one of the agents needed to continue muscular contraction and so brings about relaxatipn.

THE PERMEASE SYSTEM

In some bacteria, e.g., *E. coli*, sugar transport occurs through a special transport system known as the *permease system*. Such a system can be demonstrated most readily in cases in which it is not *a constitutive* enzyme system (one always present and demonstrable) but rather develops or is *induced* after the addition of a substrate.

The following examples serve to illustrate the phenomenon. The wild-type strain of *E. coli*, placed in, a solution of lactose, a disaccharide, does not use this sugar at once but only after a lapse of time. This delay in utilisation could be interpreted as the time required either for the entry of the sugar into the cell or for the induction of the enzyme (β-galactosidase) necessary to digest the disaccharide into monosaccharides before its use in metabolism.

It can readily be demonstrated that in wild-type *E. coli* the digestive enzyme is always present in the cell. (The cells need only be lysed and the digestive enzyme can be isolated. The time lapse in utilisation of lactose must be due then to failure of the lactose to enter the cell until some change occurs. The nature of this change has been analysed by the use of a number of mutant strains.)

A mutant of *E. colt* was isolated which apparently lacks the constitutive digestive enzyme since the enzyme could not be detected in extracts of the bacteria. As a consequence, the mutant presumably cannot use lactose in its metabolism. Yet after an induction period it is able to remove lactose from the medium, as shown by quantitative determinations before and after the induction period.

It was found that if the cell walls are removed (by lysozyme), the mutants accumulate lactose to the point that the protoplasts swell extensively and even burst, taking up water in response to the increased osmotic pressure. As much as 22 per cent of the dry weight of the bacteria consists of lactose at this time. Entry of lactose in this case is not passive but rather enzymatically controlled.

Still another mutant of *E. coli* has been found which possesses the digestive enzyme (as demonstrated in extracts of the bacteria) but is unable to use the lactose because, even after an inductive period, it is unable to develop the permease system. There is thus no doubt that separate systems are involved in uptake and in digestion of lactose.

The separateness of uptake mechanism and utilisation can be shown in yet another way, i.e., by the use of a metabolically inactive analogue such as thiomethyl-D-β-galactoside which is structurally similar to lactose. This lactose analogue is not normally utilised by wild type *E. coli,* but when supplied to the wild type divested of cell walls, the protoplasts take up the analogue to the point of swelling and even bursting.

Thus they can neither digest nor utilize the lactose analogue, but they can accumulate it-the permease system fails to distinguish between it and lactose. Even more convincing are experiments demonstrating that induction of the permease system, as that of any other induced system, is poisoned by inhibitors of protein synthesis such as chloramphenicol and by ultraviolet radiation, which inhibits various cell syntheses including enzyme inductions. It is apparent in the example just given that uptake of lactose occurs against a concentration gradient and must require work. One might therefore expect it to be blocked by metabolic poisons which generally block active transport.

Azide and dinitrophenol are indeed found to stop accumulation of nutrients in permease systems. Since these poisons prevent the formation of high energy phosphate bonds during respiration, it is possible that such bonds are involved in uptake of materials by permease systems. Monod has concluded that the permease system is a specialised enzyme transport system which, once formed, is independent of and does not participate in metabolism-it only brings the sugar (or other nutrient) through the cell membrane.

Any coupling with metabolism is, therefore, physical, not chemical. Permeases have been postulated for a variety of substrates, each one specific to a substrate. *E. coli* is thought to have 30 to 60 such systems. It is conceivable that some, if not all, are coupled to energy donors. The permease presumably combines with the substrate molecule, aiding it through the osmotic barrier.

Permeases have been described for a number of species of bacteria and for yeast. They may occur in plant and animal cells but are much more difficult to demonstrate in such materials.

Although active transport of water has been considered a possibility by some workers, others are unwilling to accept the evidence presented. True, water is voided in hypotonic urine, but filtration of the blood and active reabsorption of the solutes in the filtrate during the countercurrent flow between uriniferous tubules and the parallel small blood vessels, into which the solutes are absorbed, would account for this satisfactorily in the glomerular kidney.

However, the metabolic cost to the cell of active extrusion of the quantity of water involved alone is considered prohibitive. The various examples cited give some idea of the great variety of biological processes which are dependent upon active uptake of compounds or ions against concentration gradients. Such active uptake occurs in plant cells, animal cells and microorganisms.

It may appear surprising that cells bring in many substances by active transport when some of them might enter on the basis of their own kinetic energy, i.e., when present outside the cell in higher concentration than inside. However, active transport makes it possible for the cell to take in a particular substance at a rate determined solely by the needs of the cell at any time, regardless of how favourable or unfavourable the concentration gradient may be for its uptake at that time.

Thus, to a cell, active transport of a substance has considerable advantage over its diffusion through the cell membrane in answer to a concentration gradient, in spite of its cost in metabolic energy to the cell.

THE EFFECT OF ENVIRONMENT UPON ACTIVE TRANSPORT

Because temperature affects metabolic processes it is expected to influence the rate of active transport. If a substance is carried. through a cell membrane by active transport, the rate of liberation of energy by respiration is likely to be the limiting factor, and a temperature coefficient characteristic of thermochemical reactions is to be expected; that is, the rate of entry of such a compound should be doubled or tripled by a rise of 10°C.

However, a high temperature coefficient obtained for uptake of any substance would not by itself prove that the substance is taken in by an active process. It must be corroborated by other means as well.

Other environmental factors which affect the metabolic system are likely to affect active transport also; for example, narcotics and anesthetics, heavy metals, radiations, salts and pH. Analysis of the effects of the environment on active transport, although necessary, has not yet been extensive enough for a fruitful discussion.

POSSIBLE MECHANISM OF ACTIVE TRANSPORT

Many suggestions for the actual mechanisms of active transport, including a working model for ion transport, have appeared in the literature during recent years, but without general agreement. Among these are suggested the following:

a. A contractile protein molecule is conceived to attach to the membrane, extending and contracting (coiling). Uncoiling involves the breaking of intramolecular bonds, thus forming affinities which can be satisfied by adsorption of foreign molecules or ions.

When the protein is recoiled the original bonds are re-formed and the foreign matter desorbed inside the cell. Such protein molecules perhaps may also act as enzymes capable of liberating energy from high energy phosphate compounds by which the work necessary for active transport is accomplished.

b. Nucleic acid molecules attach themselves to incoming molecules or ions.

c. Electroendosmosis or mass movement of fluid (fluid circuit) occurs through a fixed membrane as a result of a difference in potential on both sides of it.

d. Enzyme carriers in the reduced state act as carriers and when oxidised release the ions. This has also been called electron-linked transport and the prototype is Lundegard's theory. Electrons are transferred along a chain of enzymes in the main line of oxidations as the iron in the cytochromes and cytochrome oxidase changes in valency from two to three.

Cation transport is supposed to be accomplished by the exchange of hydrogen ions for the cations in question. The source of energy is the drop in potential of the electrons. Although it is appealing, the theory is not in quantitative agreement with other findings. A variation of the enzyme transport hypothesis postulates a Mg^{++}-requiring enzyme system (an adenosinetriphosphatase) present in the plasma membrane with two closely associated sites, an outer end and an inner one.

When the enzyme transport system is activated by adenosine triphosphate the outer site has a greater affinity for potassium ion than for sodium ion, whereas the inner site has a greater affinity for sodium ion than for potassium ion. Splitting of a high energy phosphate bond from the adenosine triphosphate (with the formation of adenosine diphosphate and inorganic phosphate) reverses these affinities, as a consequence of which sodium ion is expelled from the inside of the cell while potassium ion is taken in, completing the cycle. The enzyme system is sensitive to low concentrations of the cardiac glycoside G-strophanthin (or ouabain, as it is more generally called). It requires Mg^{++} and both Na^+ and K^+ for its operation; in absence of K^+, Na^+ is not "pumped out."

Sodium was localised in the cell by electron microscopic studies as sodium pyroantimonate in ouabaintreated corneal cells. Evidence for such a system has been found in the cells of 29 of 36 tissues in the cat tested and in 10 other species. Although the quantitative details vary, the essential qualitative features remain the same.

5. Carriers other than respiratory enzymes are responsible for active transport. Consider Figure elsewhere in this chapter as an example of a generalised scheme for active transport across cells of a tubule or across a cell membrane.

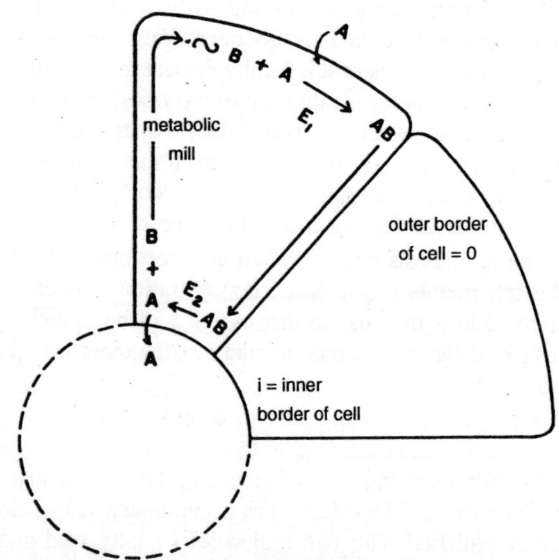

Figure 11.19: Diagram of a possible mechanism of active transport.

A cell of a nephric tubule is bathed with a solution in which a certain compound (A) is present in lesser concentration on the outside than in the lumen of the tubule. The problem is to get A through the cell and into the lumen of the tubule against the concentration gradient. Suppose that in the outer surface of the cell (0) there is present a compound with a high energy bond ($\sim$B), continually produced by metabolic reactions.

When the compounds A and ~B meet in the presence of a certain enzyme (E_1), they form a complex (AB). As the complex forms at the outer surface of the cell it diffuses through the cell down a concentration gradient. When AB reaches the inner surface (*i*) of the cell, bordering the lumen, it meets another enzyme (E_2) which decomposes it into A and B. Compound A, present in high concentration locally at the surface of the cell along the lumen of the tubule, diffuses into the lumen of the tubule. The complex AB is thus continually produced at 0 and decomposed at *i* by enzymatic reactions.

Compound B, the molecular carrier, is conceived of as a large molecule which cannot readily pass through the cell membrane; consequently it and its complex with A remain inside the cell.

This formulation has the appeal of simplicity since the complex formed at the cellular surface farthest from the lumen diffuses through the cell in a concentration gradient, and A diffuses from the cell in a concentration gradient when it is liberated from the complex at the inner surface of the cell where it is locally present in high concentration. This scheme requires polar localisation of enzymes, such as E_1 and E_2, of which there is some evidence in both kidney tissues and gut. Formation of some substances like ~B is well known in the metabolic cycles of cells. However, identification of enzymes or compounds involved in active transport has so far not met with success.

The same hypothesis might explain unidirectional active transport through the cell membrane against a concentration gradient, as in the accumulation of ions in cells. In that case 0 will be thought of as the outside surface of the cell membrane and i as the inner one, bordering on the cytoplasm.

Some attempts have been made to determine the identity of the carrier molecule (or molecules) in active transport of this type. From studies on the salt-secreting cells of glands in the eyes of marine birds (e.g., the albatross) good evidence has accumulated that phosphatidic acid (glycerol esterfied with two molecules of fatty acid and one of phosphoric acid) might be the carrier of sodium and possibly of hydrophilic molecules and ions generally.

Phosphatidic acid has the property of ready solubility in the lipid phase, while at the same time it has a minus charge which can bind the sodium. It does not leave the lipid phase but makes contact with the aqueous phase, there releasing the aqueous soluble material. The energy required for the movement is believed to be supplied by the high energy phosphate bond. From studies with other cells a protein carrier is suggested, presumably enzymatic.

PINOCYTOSIS

Edwards, in 1925, working with *Amoeba,* noticed the formation of small channels in the cell membrane followed by the separation of small vacuoles, or *pinosomes,* in the cytoplasm.

Lewis in 1931 described the same phenomenon which he saw in cells of tissue culture and gave it the name *pinocytosis*. Some more detailed observations of the process were made by Mast and Doyle on *Amoeba proteus* in 1934. Little work then appeared upon the topic until recently when workers in a number of laboratories developed an intense interest in the process.

Occurrence

The study of pinocytosis has been much facilitated by modern technological methods. For example, the development of the electron microscope has made possible resolution of much smaller channels and pinosomes. Also, some molecules, such as proteins labeled with dyes which fluoresce in the ultraviolet, when used to induce pinocytosis experimentally, can be identified readily inside the cell by fluorescence microscopy.

They can be traced, following pinocytosis, in the large number of minute vacuoles scattered in the cytoplasm. Organic molecules labeled with radioactive tracers such as ^{14}C may also be employed. In this case, autoradiographs are made of the cells and if pinocytosis has occurred, small vacuoles containing radioactive material are evident. Ferritin, a protein containing much iron has also been used. The iron, having a high density, is opaque to electrons and, following pinocytosis, electron micrographs reveal it in small vacuoles in the cytoplasm.

Thorium dioxide is also useful in the same way. It has been shown that pinocytosis occurs in ameba and ameboid cells, such as leukocytes (white blood cells), brush border cells of the kidney (active in fluid exchanges), epithelial cells of the intestine, reticuloendothelial cells, Kupffer (phagocytic) cells of the liver, cells lining the walls of the capillaries, macrophages, fibroblasts and some malignant cells in tissue culture and some other cells.

Pinocytosis was not observed (using labeled protein) in cardiac, smooth or skeletal muscle cells or in cells of a number of other tissue culture cells, including human amnion cells. In cells of plant roots, little convincing evidence has been found for pinocytosis, although labeled proteins and enzymes were found to penetrate into these cells from the outside and to appear in the cytoplasm around the nucleus. Electron micrographs of plant cells in some instances suggest canals leading

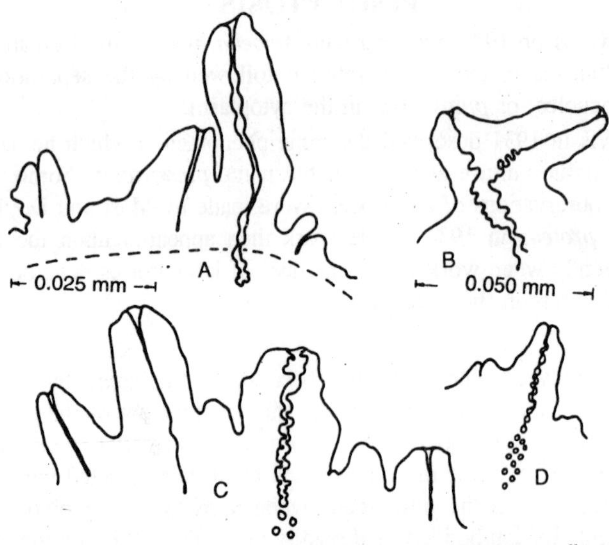

Figure 11.20 : Pinocytosis channels in Amoeba proteus: A, channels in small pseudopods in various stages of formation; B, convoluted channels; C, channel beginning to disintegrate at the inner end; D, further disintegration showing that many drops of fluid are ingested in the formation of each channel.

from the surface into the cytoplasm which might be interpreted as pinocytosis channels, but reevaluation of these results has led to the conclusion that pinocytosis is questionable in plant cells. Indirect evidence also indicates that pinocytosis probably occurs qn a wide scale in cells of embryos, since many proteins are known to pass through the placenta and immunologically specific proteins have been found to enter a wide variety of tissue cells, even into the cell nuclei.

The entry of the enzyme ribonuclease into many types of cells has also been demonstrated indirectly by its hydrolysis of the cellular RNA, as revealed by the subsequent decreased staining by basic dyes and the decreased absorption of ultraviolet radiation by the cells.

Pinocytosis has also been observed in the developing eggs of *Lymnaea stagnalis*. It is thought that pinocytosis is perhaps a general phenomenon and the negative evidence is discounted on the basis that the appropriate conditions may not yet have been supplied to the cells. The phenomenon is perhaps limited to cells capable of forming pseudopodia or at least having highly mobile membranes.

Induction of Pinocytosis

Mast and Doyle showed that in appropriately starved cells *Amoeba proteus* pinocytosis is induced by salts and proteins. It is now known that

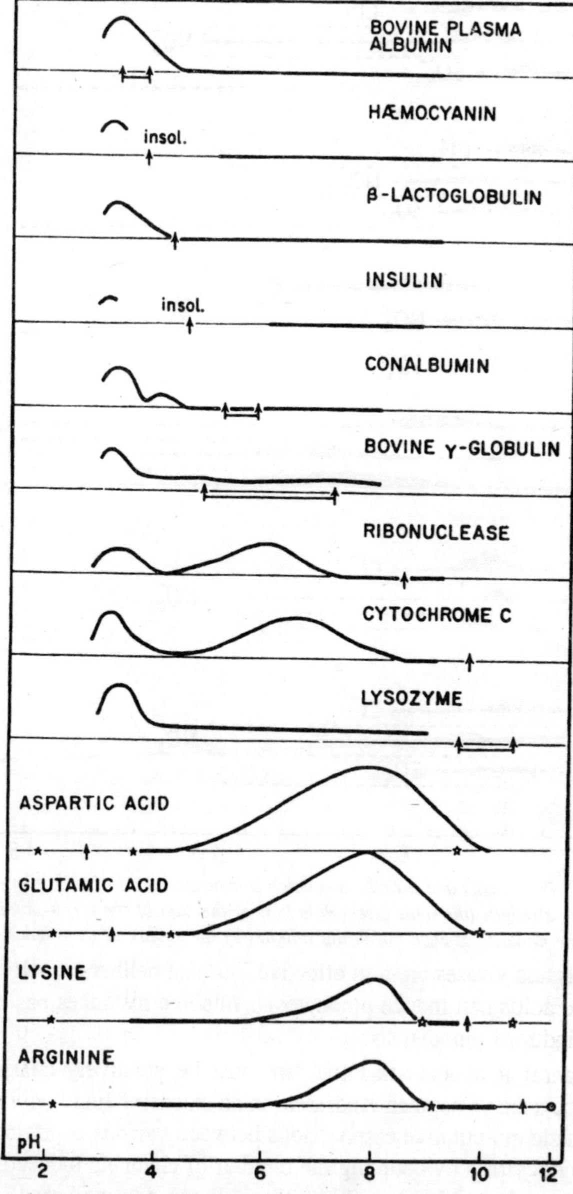

Figure 11.21: Effectiveness of induction of pinocytosis, as judged by the mean number of channels induced in 10 amebas in 20 minutes, of solutions of various amino acids (above) and proteins (below). Cysteine, asparagine, triglycine, glutamine, methionine, glycine and histidine were ineffective except for a very slight action at very high pH values.

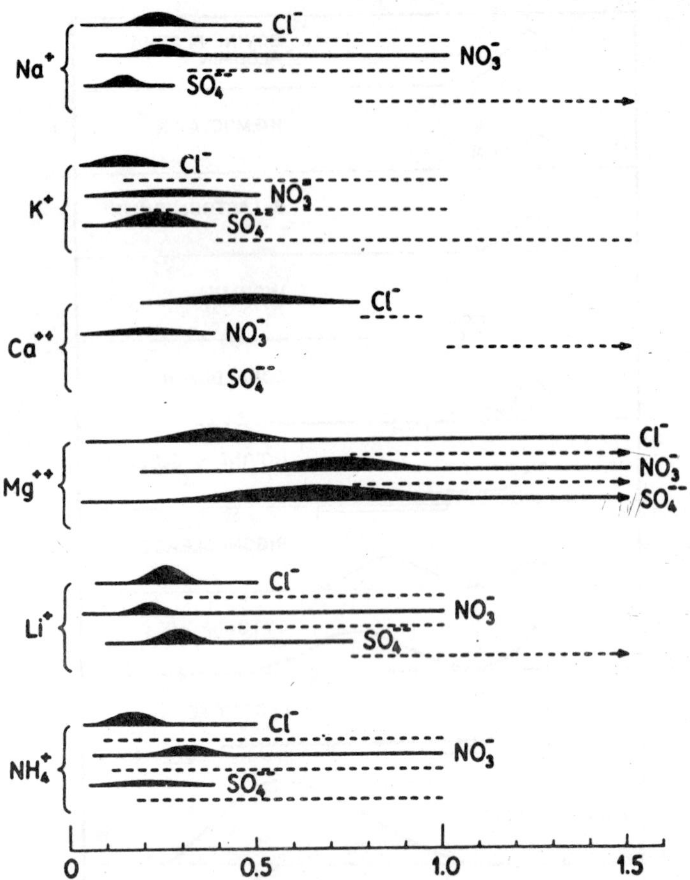

Figure 11.22: Pinocytosis and toxicity in relation to concentrations of salts expressed in terms of ionic strength (the ionic strength is half of the sum of the terms obtained by multiplying the molality of each ion in the solution by the square of its valence).

amino acids and viruses are also effective, but that neither carbohydrates nor nucleic acids can induce pinocytosis. In some instances basic dyes have also induced pinocytosis.

In general it appears that the ion must be positively charged to induce pinocytosis. Chapman-Andresen, in an extensive study on *Amoeba proteus,* made quantitative comparisons between various conditions for inducing pinocytosis by counting the number of channels induced in 10 amebas (starved two days) in 20 minutes. Some data gathered in this manner are cited from her monograph in Figures elsewhere in this chapter. The proteins found most effective for induction of pinocytosis

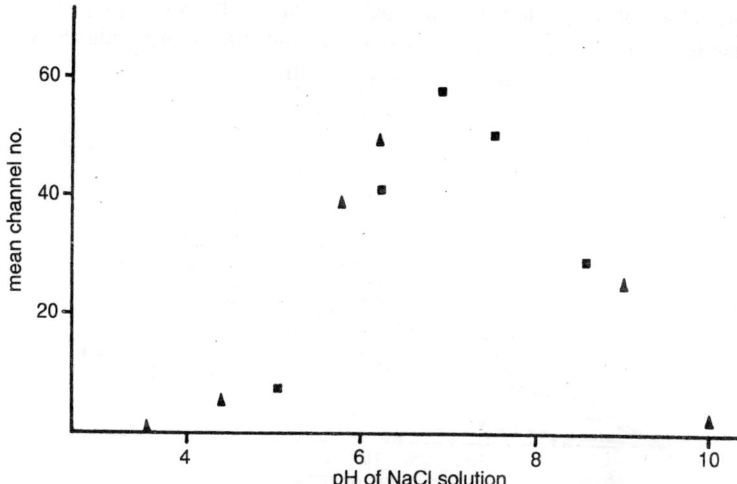

Figure 11.23: Effect of pH of inducing solution on response of Amoeba proteus to 0.125M NaCl in distilled water.

in *Amoeba* are shown in Figure elsewhere in this chapter. For leukocytes, solutions of 1 to 4 per cent albumin and globulin were effective, although even less than 1 per cent would suffice. It is necessary to set the pH of the solutions of proteins and amino acids on the acid side of the isoelectric pH, where they are positively charged, as it is only positively charged ions that are effective in inducing pinocytosis.

Salts induce pinocytosis also, cations being more effective than anions, as shown in Figure elsewhere in this chpater. Potassium, sodium and magnesium salts cause induction more readily than calcium. Sodium and magnesium induce narrower channels than potassium. A pH of 6.5 to 8.0 and a temperature between 12 and 22°C are most favourable. The optimal concentration is different for each salt, although 0.1 M solutions are adequate in most cases.

The upper limit of concentration, however, is set by the toxic action of the salt upon the cytoplasm and lies between 0.25 and 1.0 M. Since pinocytosis has been seen in the eleocytes of the sea urchin, it would be interesting if the salt relations in such a form were to be tested since the cells are already accommodated to a concentration of about 0.5 M. It was found that more pinocytotic channels were induced in enucleated than in nucleated pieces of *Amoeba proteus* (in 0.1 M NaCl) 90 minutes after cutting.

However, if the tests are made 5 days after the cutting (during which time the enucleated pieces have presumably deteriorated) many more channels are induced in the nucleate than in the enucleate pieces. Visible light has no effect on pinocytosis, but ultraviolet induces it. There also appears to be no seasonal variation.

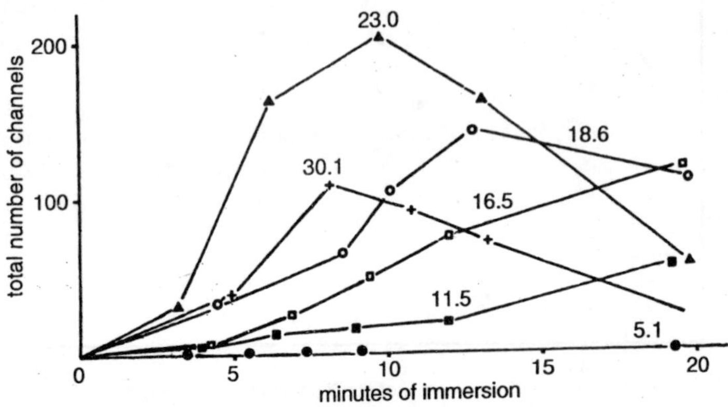

Figure 11.24: Effect of temperature on pinocytosis in Amoeba proteus in 0.125M NaCl at pH 6.5. The temperature is indicated to the side of each of the curves. No pinocytosis occurred at 5.1°C.

Once pinocytosis has been induced, in amebas, it goes on actively for 15 to 45 minutes, and then decreases. It was found necessary to wash the amebas free of ionic medium and then to reimmerse them in the salt or protein solutions in order to induce another period of activity.

Leukocytes appear to be able to continue pinocytosis for longer periods than *Amoeba.* It is presumed that only a portion of the surface of the leukocyte is active at any one time; therefore, pinocytotic action continues perhaps at all times.

Inducers of pinocytosis may cause a 50-fold decrease in electrical resistance of the plasma membrane prior to formation of typical channels and vacuoles. At this time structural changes in the unit membrane can be observed-the middle electron-transparent layer increases about twofold.

Such changes in the *Amoeba* membrane are presumably dependent upon the calcium concentration in the external medium. Pinocytosis in Sarcoma 180 II tumor cells resembles the process observed in *Amoeba.* Also, basic polyamino acid inducers are more effective the higher their molecular weight, and are ineffective if smaller than 1000.

Steric factors play a role since poly-n-lysine is more effective than poly-L-lysine. Acidic polyamino acids may be inhibitory (e.g., polyglutamic acid). Pinosomes combine with lysosomes resulting in digestion of foreign proteins.

THE PINOSOMES

When an ameba is placed into a solution favourable for induction of pinocytosis it becomes detached from the substrate, the contractile vacuole temporarily ceases operation and it forms a rosette, most of its surface being covered by small active pseudopodia, each made up almost entirely of clear ectoplasm. In the pseudopodia may be seen the clear channels which are the forerunners of the pinosomes.

By the use of proteins marked with fluorescent dyes, the pinosomes, formed by the pinching off of the vacuoles, may be readily seen inside the ameba. The pinocytotic vacuoles contain within them the same solutes as the external medium. Holier has shown that the vacuoles in an ameba do not centrifuge down soon after formation but rather stay at the centripetal end with the lighter materials of the cell. If the ameba is centrifuged after a lapse of time, the vacuoles are carried down to the centrifugal end with the heavier materials in the cell.

Presumably in the meantime the material in the vacuoles has been concentrated by the withdrawal of water. This can be shown by the use of fluorescent proteins which gradually become concentrated into a dense, brilliantly fluorescent globule. Finally, only a fluorescent granule remains when most of the water is withdrawn.

It is of interest to know whether the membrane of the pinocytotic vacuole is like the plasma membrane, and whether other materials besides water can also enter the cytoplasm from the vacuole. As a test case, glucose marked with ^{14}C was included in the inducing solution containing protein. Autoradiographs, following pinocytosis, indicated that the glucose had entered into the pinocytotic vacuoles and later autoradiographs showed that the glucose became dispersed in the cell.

Ultimately, about 15 per cent of the radioactive carbon became incorporated into the cytoskeleton of the cell and about 75 per cent of the radioactive carbon was given off by the ameba as carbon dioxide. Similar experiments carried out with methionine labeled with ^{35}S showed it to enter into the cytoplasm also.

Less has been learned about the fate of proteins taken into the cell. The fluorescein labeling on the protein is removed from the cell much more rapidly than the 14C label on glucose, but whether the protein is digested or the labeling material is just voided is not known.

Pinocytotic vacuoles have been seen to shrink, coalesce and even divide, but never to disappear. Whether the membrane of a vacuole is digested is not known. It has even been inferred that pinocytosis might be a way of building the laminated structures in the cytoplasm.

MECHANISM OF PINOCYTOSIS

Pinocytosis in the ameba, *Chaos chaos,* has been described as a result of movement of the plasma membrane, which is considered to be an elastic structure fitting closely to the underlying cytoplasm. The plasma membrane wrinkles when a flow of a pseudopod begins and the membrane is not yet freed from the gel adjoining it. This leads to invagination of parts of the plasma membrane which are consequently separated from the larger advancing portion of the membrane.

It is thought that the charged solute particles, used to induce pinocytosis, may weaken the structural rigidity of the membrane, allowing parts of it during movement to be drawn into the cytoplasm by its remaining attachments to the gel sheet. These vacuoles or channels are then separated from the membrane and released into the interior of the cell as pinosomes. Bennett on the other hand, postulated that pinocytosis does not involve just the engulfment of fluid but that there is need for attachment of materials on the surface of the cell before the vacuole can form, as shown in Figure elsewhere in this chapter.

Decisive evidence for this mechanism has been obtained, using protein as the inducer of pinocytosis. The amebas were immersed in an antigen protein for varying periods. They were then freeze-dried and sectioned. The sections were exposed to an antibody protein labeled with fluorescein. The sections so treated showed a brilliant fluorescence under ultraviolet radiation, indicating that the surface of the ameba was heavily coated with the antigen to which the antibody containing fluorescein had become attached. Further proof of attachment of protein came from studies using protein labeled with radioactive material.

In 5 minutes an ameba binds 50 times the amount of protein contained in its own volume (of solution). It would therefore appear that such binding may play an important role in pinocytosis and may be a necessary first step, so that pinocytosis is not just "cell drinking" but a process consisting of several steps.

Basic dyes (e.g., alcian blue, toluidine blue) are found to induce pinocytosis also and can be shown to be bound to the surface of an ameba, most likely by electrostatic forces. Proteins compete for the surface of an ameba in this reaction. Some photomicrographs indicate that the surface of *Amoeba proteus* has hairlike projections of extracellular material 1000 to 2000 A long and 60 A wide which greatly increase the area of the surface available for binding. The extraneous membranes, composed of mucopolysaccharides, carry a

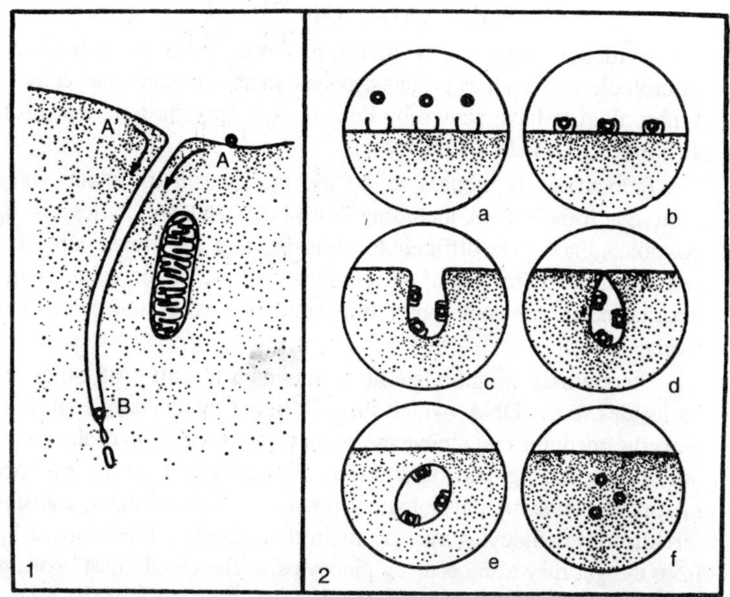

Figure 11.25: Induction of pinocytosis by attachment of particles as envisaged by Bennett. A particle (IA) induces membrane flow, forming a channel containing the particle (IB) which is taken inward into a pinosome such as those illustrated forming. The series of figures illustrates the approach (2a) and attachment (2b) of a particle, invagination of the cell membrane (2c), formation of a pinosome or pinocytotic vaculole (2a e), and, finally, incorporation of the particulate material into the cytoplasm (2J).

negative charge reacting with the cations, which act as inducers. Experiments show that to these structures, as a first step in pinocytosis, particles become attached.

Binding, or surface adsorption, of protein on the surface of the cell continues even in the presence of carbon monoxide or cyanide, either of which is inhibitory to the oxidative metabolism of the cell. It also continues at low temperatures which are inhibitory to completion of pinocytosis.

This indicates that the first step in the process of pinocytosis is a simple physical reaction. The reactions which follow, however, including the formation of channels and vesicles, are stopped by metabolic inhibitors in *Amoeba* (but not in sarcoma cells) and also by low temperatures.

This indicates that they are dependent upon the oxidative metabolism of the cell. Thus it seems that there are at least two kinds of processes involved in pinocytosis.

SIGNIFICANCE OF PINOCYTOSIS

Evidence from many different kinds of cells indicates that macromolecules such as proteins, polypeptides and hormones enter cells. Embryologists have shown by delicate and specific tests that proteins also enter into cells of embryos.

Pinocytosis is thought to be the mechanism by which entry of such macromolecules including hormones and antibodies, is made possible, since it is difficult to consider that the small pores in the cell membrane (measured in a number of ways) or solution in the plasma membrane could possibly permit such macromolecules to enter freely.

Pinocytosis might also be a means for entry of other large molecules, e.g., DNA, which the cell might incidentally engulf along with the medium containing the pinocytosis-inciting particles or ions, possibly inducing transformation. Pinocytosis may be of special importance in some physiological processes. For example, transfer of ferritin in reticulocytes may occur in this manner. Transport of lipids from the gut may also occur by pinocytosis. The blood "dust" consisting of fine lipid droplets may well have been liberated into the blood by the reverse of pinocytosis.

Is Pinocytosis Active Transport?

If active transport is defined as the selective uptake of a particular solute against a concentration gradient performed at the expense of metabolic energy, then pinocytosis does not truly qualify as active transport. In pinocytosis the active step requiring energy is the formation of a vacuole or channel and its separation from the cell membrane.

This phase of the process is not selective because it is a random sampling of the environmental fluid. The only selective phase appears to be the binding of proteins and perhaps other molecules to the surface of the cell and this may or may not require metabolic energy. Perhaps surface binding is prerequisite to any active transport.

When the vacuole is formed these materials are included and they may move passively against a concentration gradient. It might even be that the entering proteins differ from those inside the cell, in which case they therefore move along with a concentration gradient, not against it.

Solutes other than the particles inducing pinocytosis, included in the engulfed vacuole as a random sample of the environmental fluid, are concentrated only as a result of subsequent processes, possibly active ones, not yet studied. Whether these solutes pass by diffusion or by active transport through the portion of the plasma membrane which forms the vacuolar wall, and whether the solutes are concentrated by removal of the water from the vacuole by osmosis or active transport is not known, but some solutes are readily incorporated into the cytoskeleton or used in metabolism, e.g., 14C glucose, mentioned before.

Some cells also void secretory vacuoles or granules -presumably by the reverse of pinocytosis. It must be recognised that pinocytosis appears to be more complex than active uptake through the plasma membrane.

INDEX

D

E

F

G

H

I

J